T0372359

Numerical Solutions of Boundary Value Problems of Non-linear Differential Equations

Numerical Solutions of Boundary Value Problems of Non-linear Differential Equations

Sujaul Chowdhury
Department of Physics, Shahjalal University of Science and Technology, Sylhet, Bangladesh

Syed Badiuzzaman Faruque
Department of Physics, Shahjalal University of Science and Technology, Sylhet, Bangladesh

Ponkog Kumar Das
Department of Physics, Shahjalal University of Science and Technology, Sylhet, Bangladesh

CRC Press is an imprint of the
Taylor & Francis Group, an **informa** business

First edition published 2022
by CRC Press
6000 Broken Sound Parkway NW, Suite 300, Boca Raton, FL 33487–2742

and by CRC Press
2 Park Square, Milton Park, Abingdon, Oxon, OX14 4RN

Library of Congress Cataloging-in-Publication Data
Names: Chowdhury, Sujaul, author. | Faruque, Syed Badiuzzaman, author. | Das, Ponkog Kumar, author.
Title: Numerical solutions of boundary value problems of non-linear differential equations / Sujaul Chowdhury, Department of Physics, Shahjalal University of Science and Technology, Sylhet, Bangladesh, Syed Badiuzzaman Faruque, Department of Physics, Shahjalal University of Science and Technology, Sylhet, Bangladesh, Ponkog Kumar Das, Department of Physics, Shahjalal University of Science and Technology, Sylhet, Bangladesh.
Description: First edition. | Boca Raton : Chapman & Hall/CRC Press, [2022] | Includes bibliographical references and index. | Summary: "Numerical Solutions of Boundary Value Problems of Non-linear Differential Equations presents in comprehensive detail numerical solution to boundary value problems of a number of non-linear differential equations. Numerical solutions have been presented in comprehensive detail Newton's iterative method has been applied to solve system of non-linear algebraic equations encountered In each case, Euler solutions have been obtained to serve as a cross-check as to any mistakes Mathematica has been used as the program. Programs written in Mathematica have been presented for re-use This book is primarily aimed at final year undergraduate students of Physics and Mathematics who have undertaken a course on computational physics"—Provided by publisher.
Identifiers: LCCN 2021032684 (print) | LCCN 2021032685 (ebook) | ISBN 9781032069951 (hardback) | ISBN 9781032149295 (paperback) | ISBN 9781003204916 (ebook)
Subjects: LCSH: Boundary value problems—Numerical solutions. | Nonlinear boundary value problems—Numerical solutions.
Classification: LCC QC20.7.B6 C46 2022 (print) | LCC QC20.7.B6 (ebook) | DDC 515/.35—dc23
LC record available at https://lccn.loc.gov/2021032684
LC ebook record available at https://lccn.loc.gov/2021032685

ISBN: 978-1-032-06995-1 (hbk)
ISBN: 978-1-032-14929-5 (pbk)
ISBN: 978-1-003-20491-6 (ebk)

DOI: 10.1201/9781003204916

Typeset in Times
by Apex CoVantage, LLC

CONTENTS

PREFACE

This book, titled *Numerical Solutions of Boundary Value Problems of Non-linear Differential Equations*, is my fifth book on numerical methods for Computational Physics using the user-friendly program Mathematica. Here, boundary value problems of a number of non-linear differential equations with increasing difficulty have been solved and solutions have been presented in comprehensive detail. The differential equations are $\frac{dy}{dx}+y^2=x$ $\frac{dy}{dx}+y^2=Sin(x)$ $\frac{dy}{dx}+y^2=Sin^2(x)$ $\frac{dy}{dx}+y^3=x$ $\frac{d^2y}{dx^2}+y^2=x$ $\frac{d^2y}{dx^2}+y^2=x^2$ and $\frac{d^2y}{dx^2}=-\frac{y}{(y^2+r^2(x))^{3/2}}$ where $r(x)=-c\,Cos(x)+\frac{1}{2}c^2(1-Cos(2x))$

where c is a constant. A complete and self-contained introduction is covered in the first chapter so that readers need not consult other books to understand the methods. In the remaining seven chapters, we took up the seven differential equations one by one. In each case, we first cracked the problem using the Euler's method. This enabled us to get two boundary values which are crucial for the rest of the work. We began the main work by replacing derivatives with finite difference approximations in the differential equations. This led to a system of non-linear algebraic equations which we have solved using Newton's iterative method as applied to a system of non-linear algebraic equations and obtained solution to the differential equation. In each case, we have ascertained that the iterations converge to Euler solutions. Agreement between results of Newton's method and that of Euler's method served as a cross-check for mistakes. No noteworthy differences were found. To our surprise, we discovered that except for the boundary values, initial values of the 1st iteration need not be anything close to the final convergent values of the numerical solution. In each chapter, programs written in Mathematica 6.0 have been presented for re-use; numerical solutions have been presented as tabulated set of values as well as graphically. Microsoft Excel was used to obtain the magnificent plots. The work presented in this book explores the use of Newton's method in obtaining numerical solutions to boundary value problems of non-linear differential equations. The book is a good read and will contribute to applied mathematics and physics. Keywords for the book are: numerical solution, boundary value problem, non-linear differential equation, finite difference method, Newton's iterative method, system of non-linear algebraic equations, Euler solution. Chapter 8 would not be taken up without Syed Badiuzzaman Faruque and would not be completed without the help of Ponkog Kumar Das.

Sujaul Chowdhury
Sylhet, Bangladesh
17 February 2020

AUTHOR BIOGRAPHY

Sujaul Chowdhury is Professor in the Department of Physics, Shahjalal University of Science and Technology (SUST), Bangladesh. He obtained a BSc (Honors) in Physics in 1994 and MSc in Physics in 1996 from SUST. He obtained a PhD in Physics from the University of Glasgow, UK in 2001. He was a Humboldt Research Fellow for one year at the Max Planck Institute, Stuttgart, Germany.

Syed Badiuzzaman Faruque is Professor in the Department of Physics, SUST. He is a researcher with interests in Quantum Theory, Gravitational Physics, and Materials Science. He has been teaching Physics at university level for about 27 years. He studied Physics at the University of Dhaka, Bangladesh and at the University of Massachusetts Dartmouth, USA and obtained a PhD from SUST.

Ponkog Kumar Das is Assistant Professor in the Department of Physics, SUST. He obtained a BSc (Honors) and MSc in Physics from SUST. He is a promising future intellectual.

1

INTRODUCTION

In this chapter, we have narrated the problem and the methodology for the entire book.

1.1 THE NON-LINEAR DIFFERENTIAL EQUATIONS WE SOLVED IN THIS BOOK

We have numerically solved boundary value problems of the following non-linear differential equations:

$$\frac{dy}{dx} + y^2 = f(x) \tag{1.1}$$

$$\frac{dy}{dx} + y^3 = f(x) \tag{1.2}$$

$$\frac{d^2y}{dx^2} + y^2 = f(x) \tag{1.3}$$

$$\frac{d^2y}{dx^2} = -\frac{y}{(y^2 + r^2(x))^{3/2}} \tag{1.4}$$

where

$$r = -c \ \text{Cos}(x) + \frac{1}{2}c^2(1 - \text{Cos}(2x)) \tag{1.5}$$

where c is a constant parameter and $f(x)$'s are functions of x. That is, we have obtained tabulated set of values (x_i, y_i) in the interval $x = a$ to b as numerical solutions of these differential equations, provided values of y for $x = a$ and $x = b$ are known or given. Equation (1.4) comes from reference [1].

1.2 APPROXIMATION TO DERIVATIVES

Let us divide the interval $x = a$ to b in n equal parts, each part being h. As such, $x_0 = a$, $x_1 = a + h$, $x_2 = a + 2h$, $x_3 = a + 3h$, . . ., $x_i = a + ih$, . . ., $x_n = b$. Let y_i be the value of y for $x = x_i$.

DOI: 10.1201/9781003204916-1

With e.g. polynomial interpolation in mind, we can regard y as a continuous function of x. As such, Taylor's series gives us

$$y(x+h) = y(x) + h\,y'(x) \tag{1.6}$$

if y is a slowly varying function of x. More accurately, we have

$$y(x+h) = y(x) + h\,y'(x) + \frac{h^2}{2}\,y''(x) \tag{1.7}$$

Again, we have

$$y(x-h) = y(x) - h\,y'(x) \tag{1.8}$$

and

$$y(x-h) = y(x) - h\,y'(x) + \frac{h^2}{2}\,y''(x) \tag{1.9}$$

Equations (1.6) and (1.8) give

$$y'(x) = [y(x+h) - y(x-h)]/(2h) \tag{1.10}$$

Equations (1.7) and (1.9) give

$$y(x+h) + y(x-h) = 2\,y(x) + h^2\,y''(x)$$

or,

$$y''(x) = [y(x-h) - 2\,y(x) + y(x+h)]/h^2 \tag{1.11}$$

Equation (1.10) gives

$$y' = [y_{i+1} - y_{i-1}]/(2h) \tag{1.12}$$

and equation (1.11) gives

$$y'' = (y_{i-1} - 2\,y_i + y_{i+1})/h^2 \tag{1.13}$$

Equations (1.12) and (1.13) are called *finite difference approximation* to derivatives.

1.3 STATEMENT OF THE PROBLEM

Suppose we do not know analytical solutions of differential equations (1.1) or (1.2) or (1.3) or (1.4). But we somehow know values of y for

$x = a$ and $x = b$. If we replace the derivatives dy/dx or d^2y/dx^2 by *finite difference approximations* given by equations (1.12) or (1.13) and y by y_i in differential equations (1.1) or (1.2) or (1.3) or (1.4), we end up with a set or system of non-linear algebraic equations. We have solved the system of non-linear algebraic equations by Newton's method as applied to a system of non-linear equations, and we have got tabulated set of values of (x_i, y_i) as numerical solutions of differential equations (1.1) or (1.2) or (1.3) or (1.4).

We have also obtained Euler solutions (x_i, y_i) of differential equations (1.1) or (1.2) or (1.3) or (1.4) and compared with the results of the Newton's method. We have used Mathematica 6.0 in the calculations.

1.4 EULER SOLUTION OF DIFFERENTIAL EQUATION

Euler's method gives numerical solution of differential equation in the form of a set of tabulated values. Suppose we wish to solve the differential equation

$$\frac{dy}{dx} = f(x, y) \tag{1.14}$$

with known initial condition $y\,(x = x_0) = y_0$. Here f, in general, depends on both x and y.

Suppose that we wish to solve equation (1.14) to obtain values of y at

$$x = x_r = x_0 + r\,h \tag{1.15}$$

where $r = 1, 2, 3, \ldots$ and h is a constant increment of x determining the series of values of x.

Integrating equation (1.14), for x in the interval x_0 to x_1, we get

$$y_1 = y_0 + \int_{x_0}^{x_1} f(x, y)\, dx \tag{1.16}$$

Here y_1 is the value of y at $x = x_0 + h$. Taking the value of f in the interval $x_0 < x < x_1$, as an approximation, as $f(x_0, y_0)$ allows us to write equation (1.16) as

$$y_1 = y_0 + h\; f(x_0, y_0) \tag{1.17}$$

Integrating equation (1.14), for x in the interval x_1 to x_2, we get

$$y_2 = y_1 + \int_{x_1}^{x_2} f(x, y)\, dx \tag{1.18}$$

Here y_2 is the value of y at $x = x_0 + 2h$. Taking the value of f in the interval $x_1 < x < x_2$, as an approximation, as $f(x_1, y_1)$ allows us to write equation (1.18) as

$$y_2 = y_1 + h\ f(x_1, y_1) \tag{1.19}$$

Similarly, we get

$$y_{n+1} = y_n + hf(x_n, y_n) \tag{1.20}$$

Here y_n and y_{n+1} are values of y at $x = x_n$ and x_{n+1}, respectively. Equation (1.20) is the renowned *Euler formula* or *Euler solution* of differential equation (1.14). We need to use small values of the parameter h to get sets of values of (x_n, y_n) that accurately match with known exact results.

1.5 NEWTON'S METHOD OF SOLVING SYSTEM OF NON-LINEAR EQUATIONS

Suppose we have a system of n non-linear algebraic equations:

$$\begin{aligned}
&f_1(y_1, y_2, y_3, \ldots, y_n) = 0\\
&f_2(y_1, y_2, y_3, \ldots, y_n) = 0\\
&f_3(y_1, y_2, y_3, \ldots, y_n) = 0\\
&\ldots\\
&f_n(y_1, y_2, y_3, \ldots, y_n) = 0
\end{aligned} \tag{1.21}$$

where f_i 's are functions of $y_1, y_2, y_3, \ldots, y_n$.

Suppose $(y_{1m}, y_{2m}, y_{3m}, \ldots, y_{nm})$ is an approximate solution to equation (1.21) and $(\beta_1, \beta_2, \beta_3, \ldots, \beta_n)$ is the accurate solution. As such, Taylor series gives us

$$f_1(\beta_1, \beta_2, \beta_3, \ldots, \beta_n) = 0 = f_1(y_{1m}, y_{2m}, y_{3m}, \ldots, y_{nm}) + (\beta_1 - y_{1m}) f_{1y1} + (\beta_2 - y_{2m}) f_{1y2} + (\beta_3 - y_{3m}) f_{1y3} + \cdots + (\beta_n - y_{nm}) f_{1yn}$$

$$f_2(\beta_1, \beta_2, \beta_3, \ldots, \beta_n) = 0 = f_2(y_{1m}, y_{2m}, y_{3m}, \ldots, y_{nm}) + (\beta_1 - y_{1m}) f_{2y1} + (\beta_2 - y_{2m}) f_{2y2} + (\beta_3 - y_{3m}) f_{2y3} + \cdots + (\beta_n - y_{nm}) f_{2yn}$$

$$f_3(\beta_1, \beta_2, \beta_3, \ldots, \beta_n) = 0 = f_3(y_{1m}, y_{2m}, y_{3m}, \ldots, y_{nm}) + (\beta_1 - y_{1m}) f_{3y1} + (\beta_2 - y_{2m}) f_{3y2} + (\beta_3 - y_{3m}) f_{3y3} + \cdots + (\beta_n - y_{nm}) f_{3yn}$$

$$\ldots$$

$$f_n(\beta_1, \beta_2, \beta_3, \ldots, \beta_n) = 0 = f_n(y_{1m}, y_{2m}, y_{3m}, \ldots, y_{nm})$$
$$+ (\beta_1 - y_{1m}) f_{ny1} + (\beta_2 - y_{2m}) f_{ny2} + (\beta_3 - y_{3m}) f_{ny3} + \cdots + (\beta_n - y_{nm}) f_{nyn} \quad (1.22)$$

where f_{ny1} stands for derivative of f_n with respect to (w.r.t.) y_1, f_{ny2} stands for derivative of f_n w.r.t. y_2 and so on. Since terms with higher order derivatives in Taylor's series have been omitted in equation (1.22), values of $\beta_1, \beta_2, \beta_3, \ldots, \beta_n$ obtainable from equation (1.22) are not exact solutions of equation (1.21), but it will usually be a better approximation than $(y_{1m}, y_{2m}, y_{3m}, \ldots, y_{nm})$. As such, we replace $\beta_1, \beta_2, \beta_3, \ldots, \beta_n$ in equation (1.22) with $(y_{1m+1}, y_{2m+1}, y_{3m+1}, \ldots, y_{nm+1})$ and get the following iterative scheme:

$$0 = f_1(y_{1m}, y_{2m}, y_{3m}, \ldots, y_{nm}) + (y_{1m+1} - y_{1m}) f_{1y1} + (y_{2m+1} - y_{2m}) f_{1y2}$$
$$+ (y_{3m+1} - y_{3m}) f_{1y3} + \cdots + (y_{nm+1} - y_{nm}) f_{1yn}$$
$$0 = f_2(y_{1m}, y_{2m}, y_{3m}, \ldots, y_{nm}) + (y_{1m+1} - y_{1m}) f_{2y1} + (y_{2m+1} - y_{2m}) f_{2y2}$$
$$+ (y_{3m+1} - y_{3m}) f_{2y3} + \cdots + (y_{nm+1} - y_{nm}) f_{2yn}$$
$$0 = f_3(y_{1m}, y_{2m}, y_{3m}, \ldots, y_{nm}) + (y_{1m+1} - y_{1m}) f_{3y1} + (y_{2m+1} - y_{2m}) f_{3y2}$$
$$+ (y_{3m+1} - y_{3m}) f_{3y3} + \cdots + (y_{nm+1} - y_{nm}) f_{3yn}$$
$$\ldots$$
$$0 = f_n(y_{1m}, y_{2m}, y_{3m}, \ldots, y_{nm}) + (y_{1m+1} - y_{1m}) f_{ny1} + (y_{2m+1} - y_{2m}) f_{ny2}$$
$$+ (y_{3m+1} - y_{3m}) f_{ny3} + \cdots + (y_{nm+1} - y_{nm}) f_{nyn} \quad (1.23)$$

Denoting $y_{1m+1} - y_{1m}$ by u_{1m}, $y_{2m+1} - y_{2m}$ by u_{2m} and so on, equation (1.23) becomes a system of algebraic equations linear in u_{nm}'s, given by

$$f_{1y1}\, u_{1m} + f_{1y2}\, u_{2m} + f_{1y3}\, u_{3m} + \cdots + f_{1yn}\, u_{nm} = -f_1(y_{1m}, y_{2m}, y_{3m}, \ldots, y_{nm})$$
$$f_{2y1}\, u_{1m} + f_{2y2}\, u_{2m} + f_{2y3}\, u_{3m} + \cdots + f_{2yn}\, u_{nm} = -f_2(y_{1m}, y_{2m}, y_{3m}, \ldots, y_{nm})$$
$$f_{3y1}\, u_{1m} + f_{3y2}\, u_{2m} + f_{3y3}\, u_{3m} + \cdots + f_{3yn}\, u_{nm} = -f_3(y_{1m}, y_{2m}, y_{3m}, \ldots, y_{nm})$$
$$\ldots$$
$$f_{ny1}\, u_{1m} + f_{ny2}\, u_{2m} + f_{ny3}\, u_{3m} + \cdots + f_{nyn}\, u_{nm} = -f_n(y_{1m}, y_{2m}, y_{3m}, \ldots, y_{nm}) \quad (1.24)$$

If we solve the system of linear equations (1.24), we get u_{nm}'s from which we can get the next iterate y_{nm+1} using

$$y_{nm+1} = y_{nm} + u_{nm} \quad (1.25)$$

which we can re-use in equation (1.24) and get the next iterate using equation (1.25). The iterative scheme narrated here is available from reference [2].

We have used Mathematica 6.0 in solving linear system of equations (1.24). Here is an introductory program to solve a system of three linear equations given by

$$
\begin{aligned}
10x + 2y + 3z &= 10 \\
x - 2y + 5z &= 5 \\
5x + 6y + 10z &= 7
\end{aligned}
$$

The command or program is

```
G=NSolve[{10x+2y+3z == 10, x-2y+5z ==5,
5x+6y+10z==7},{x,y,z}]
f[1]=EL1=Part[G,1];L1=x/.EL1
f[2]=EL2=Part[G,1];L2=y/.EL2
f[3]=EL3=Part[G,1];L3=z/.EL3
```

with the result f[1] = x = 0.9431, f[2] = y = −0.5829, f[3] = z = 0.5781.

Extensive illustrations of this are available in reference[3].

2

NUMERICAL SOLUTION OF BOUNDARY VALUE PROBLEM OF NON-LINEAR DIFFERENTIAL EQUATION

Example I

In this chapter, we have presented the numerical solution to boundary value problem of the 1st non-linear differential equation in this book in comprehensive detail.

2.1 THE 1ST NON-LINEAR DIFFERENTIAL EQUATION IN THIS BOOK: EULER SOLUTION

In this chapter, we deal with the boundary value problem of the following non-linear differential equation:

$$\frac{dy}{dx} + y^2 = x \tag{2.1}$$

We first turn to Euler solution (see e.g. reference[4], [5], [6]) of this differential equation. To this end, we re-write equation (2.1) as

$$\frac{dy}{dx} = x - y^2 \tag{2.2}$$

which looks like equation (1.14): $dy / dx = f\left(x, y\right)$ which admits the iteration given by equation (1.20): $y_{n+1} = y_n + hf(x, y)$.

We have written program number 2.1 following Euler's method and solved equation (2.2) with initial values $(x, y) = (0, 3)$ with the numerical result shown in Table 2.1. From Table 2.1, we gather the boundary values $(x, y) = (0, 3)$ and $(4, 1.9587)$.

DOI: 10.1201/9781003204916-2

TABLE 2.1
Euler Solution of Equation (2.2): $\frac{dy}{dx} = x - y^2$ with Initial Condition $(x, y) = (0, 3)$

i	x	y
1	0.1	2.1100
2	0.2	1.6848
3	0.3	1.4309
4	0.4	1.2662
5	0.5	1.1559
6	0.6	1.0823
7	0.7	1.0351
8	0.8	1.0080
9	0.9	0.9964
10	1.0	0.9971
11	1.1	1.0077
12	1.2	1.0261
13	1.3	1.0508
14	1.4	1.0804
15	1.5	1.1137
16	1.6	1.1497
17	1.7	1.1875
18	1.8	1.2265
19	1.9	1.2661
20	2.0	1.3058
21	2.1	1.3453
22	2.2	1.3843
23	2.3	1.4227
24	2.4	1.4603
25	2.5	1.4970
26	2.6	1.5329
27	2.7	1.5679
28	2.8	1.6021
29	2.9	1.6354
30	3.0	1.6680
31	3.1	1.6998
32	3.2	1.7308
33	3.3	1.7613

(continued)

i	x	y
34	3.4	1.7911
35	3.5	1.8203
36	3.6	1.8489
37	3.7	1.8771
38	3.8	1.9047
39	3.9	1.9319
40	4.0	1.9587

Program Number 2.1

```
h=0.1;
i=0;
x=0;
y=3;
Table[{i=i+1,x=x+h,y=y+h*(x-y^2)},{x,0,4,h}];
TableForm[%,TableSpacing->{2,2},TableHeadings
->{None,{"i","x","y"}}]

i=0;
x=0;
y=3;
p1=ListPlot[Table[{i=i+1;x=x+h,y=y+h*(x-y^2)},
{x,0,4,h}],
Frame->True,FrameLabel->{"x","y"},FrameTicks
->All,PlotStyle->{Black}]
```

2.2 THE 1ST NON-LINEAR DIFFERENTIAL EQUATION IN THIS BOOK: SOLUTION BY NEWTON'S ITERATIVE METHOD

For now we assume that the Euler solution is unknown and we have the non-linear differential equation (2.1): $\frac{dy}{dx}+y^2=x$ with the boundary values $(x, y) = (0, 3)$ and $(4, 1.9587)$.

We proceed by replacing the derivative dy/dx with finite difference approximation given by equation (1.12): $y' = (y_{i+1} - y_{i-1})/(2h)$. This results in the following difference equation:

$$\frac{y_{i+1}-y_{i-1}}{2h}+y_i^2=x_i \text{ or, } -y_{i-1}+2h(y_i^2-x_i)+y_{i+1}=0 \qquad (2.3)$$

We divide the interval $x = 0$ to 4 into 40 equal parts each being $h = 0.1$. We write equation (2.3) for $i = 1, 2, 3, \ldots, 39$ for which $x_i = 0.1, 0.2, 0.3, \ldots, 3.9$, respectively. We get the following system of non-linear algebraic equations:

$$
\begin{aligned}
f_1 &= -y_0 + 2h(y_1^2 - x_1) + y_2 = 0 \\
f_2 &= -y_1 + 2h(y_2^2 - x_2) + y_3 = 0 \\
f_3 &= -y_2 + 2h(y_3^2 - x_3) + y_4 = 0 \\
&\ldots \\
f_{39} &= -y_{38} + 2h(y_{39}^2 - x_{39}) + y_{40} = 0
\end{aligned}
$$

Here $y_0 = 3$ and $y_{40} = 1.9587$ are constants. As such, the corresponding system of linear equations following Newton's iterative method is as follows:

$$
\begin{aligned}
4hy_1u_1 + 1u_2 &= -(-y_0 + 2h(y_1^2 - x_1) + y_2) \\
-1u_1 + 4hy_2u_2 + 1u_3 &= -(-y_1 + 2h(y_2^2 - x_2) + y_3) \\
-1u_2 + 4hy_3u_3 + 1u_4 &= -(-y_2 + 2h(y_3^2 - x_3) + y_4) \\
&\ldots \\
-1u_{38} + 4hy_{39}u_{39} &= -(-y_{38} + 2h(y_{39}^2 - x_{39}) + y_{40})
\end{aligned}
$$

To solve these equations which are linear in u_i's, we have taken initial set of values of y[i]'s as

```
y[0]=3;y[1]=3;y[2]=2.5;y[3]=2;y[4]=2;y[5]=1.5;
y[6]=1;y[7]=2;y[8]=1;y[9]=1;y[10]=1;
y[11]=1;y[12]=1;y[13]=1;y[14]=1;y[15]=1;y[16]=1;
y[17]=1;y[18]=1;y[19]=1;y[20]=1;
y[21]=0.5;y[22]=0.5;y[23]=2.5;y[24]=0.5;y[25]=0.5;
y[26]=2.5;y[27]=0.5;y[28]=2.5;
y[29]=0.5;y[30]=0.5;y[31]=0.5;y[32]=0.5;y[33]=2.5;
y[34]=2.5;y[35]=3.5;y[36]=3.5;
y[37]=4;y[38]=3.5;y[39]=4;y[40]=1.9587;
```

in program number 2.2. In using program number 2.2, the content of the program between the lines marked as

```
aaaaaaaaaaaaaaaaaaaaaaaaaaaaaaa
```

and

bbbbbbbbbbbbbbbbbbbbbbbbbbbbbbbbbbb

should be pasted $p - 1$ times after the line

bbbbbbbbbbbbbbbbbbbbbbbbbbbbbbbbbbb

where p (= 4 in this case) is the number of iterations required to produce convergence of the values of y[i]'s. **The results of the program are shown in Table 2.2 and Figure 2.1.**

Program Number 2.2

```
h=0.1;
y[0]=3;
y[40]=1.9587;

y[1]=3;
y[2]=2.5;
y[3]=2;y[4]=2;y[5]=1.5;y[6]=1;y[7]=2;y[8]=1;
y[9]=1;y[10]=1;
y[11]=1;y[12]=1;y[13]=1;y[14]=1;y[15]=1;y[16]=1;
y[17]=1;y[18]=1;y[19]=1;y[20]=1;
y[21]=0.5;y[22]=0.5;y[23]=2.5;y[24]=0.5;y[25]=0.5;
y[26]=2.5;y[27]=0.5;y[28]=2.5;
y[29]=0.5;y[30]=0.5;y[31]=0.5;y[32]=0.5;y[33]=2.5;
y[34]=2.5;y[35]=3.5;y[36]=3.5;
y[37]=4;y[38]=3.5;y[39]=4;

x=0;
i=0;
Table[{i=i+1,x=x+h,y[i]},{i,0,38,1}];
TableForm[%,TableSpacing->{3,3},TableHeadings
->{None,{"i","x","y"}}]
x=0;
i=0;
p1=ListPlot[Table[{{i=i+1;x=x+h,y[i]}},{i,0,38,1}],
Frame->True,
FrameLabel->{"x","y"},FrameTicks->All,PlotStyle->
{Black},PlotRange->Automatic]

aaaaaaaaaaaaaaaaaaaaaaaaaaaaaaa
G=NSolve[{
```

```
4*h*y[1]*u1+1*u2==-(-y[0]+2*h*(y[1]^2-0.1)+y[2]),
-1*u1+4*h*y[2]*u2+1*u3==-(-y[1]+2*h*(y[2]^2-
0.2)+y[3]),
-1*u2+4*h*y[3]*u3+1*u4==-(-y[2]+2*h*(y[3]^2-
0.3)+y[4]),
-1*u3+4*h*y[4]*u4+1*u5==-(-y[3]+2*h*(y[4]^2-
0.4)+y[5]),
-1*u4+4*h*y[5]*u5+1*u6==-(-y[4]+2*h*(y[5]^2-
0.5)+y[6]),
-1*u5+4*h*y[6]*u6+1*u7==-(-y[5]+2*h*(y[6]^2-
0.6)+y[7]),
-1*u6+4*h*y[7]*u7+1*u8==-(-y[6]+2*h*(y[7]^2-
0.7)+y[8]),
-1*u7+4*h*y[8]*u8+1*u9==-(-y[7]+2*h*(y[8]^2-
0.8)+y[9]),
-1*u8+4*h*y[9]*u9+1*u10==-(-y[8]+2*h*(y[9]^2-
0.9)+y[10]),
-1*u9+4*h*y[10]*u10+1*u11==-(-y[9]+2*h*(y[10]^2-
1.0)+y[11]),
-1*u10+4*h*y[11]*u11+1*u12==-(-y[10]+2*h*(y[11]^2-
1.1)+y[12]),
-1*u11+4*h*y[12]*u12+1*u13==-(-y[11]+2*h*(y[12]^2-
1.2)+y[13]),
-1*u12+4*h*y[13]*u13+1*u14==-(-y[12]+2*h*(y[13]^2-
1.3)+y[14]),
-1*u13+4*h*y[14]*u14+1*u15==-(-y[13]+2*h*(y[14]^2-
1.4)+y[15]),
-1*u14+4*h*y[15]*u15+1*u16==-(-y[14]+2*h*(y[15]^2-
1.5)+y[16]),
-1*u15+4*h*y[16]*u16+1*u17==-(-y[15]+2*h*(y[16]^2-
1.6)+y[17]),
-1*u16+4*h*y[17]*u17+1*u18==-(-y[16]+2*h*(y[17]^2-
1.7)+y[18]),
-1*u17+4*h*y[18]*u18+1*u19==-(-y[17]+2*h*(y[18]^2-
1.8)+y[19]),
-1*u18+4*h*y[19]*u19+1*u20==-(-y[18]+2*h*(y[19]^2-
1.9)+y[20]),
-1*u19+4*h*y[20]*u20+1*u21==-(-y[19]+2*h*(y[20]^2-
2.0)+y[21]),
-1*u20+4*h*y[21]*u21+1*u22==-(-y[20]+2*h*(y[21]^2-
2.1)+y[22]),
-1*u21+4*h*y[22]*u22+1*u23==-(-y[21]+2*h*(y[22]^2-
2.2)+y[23]),
-1*u22+4*h*y[23]*u23+1*u24==-(-y[22]+2*h*(y[23]^2-
2.3)+y[24]),
```

```
-1*u23+4*h*y[24]*u24+1*u25==-(-y[23]+2*h*(y[24]^2-
2.4)+y[25]),
-1*u24+4*h*y[25]*u25+1*u26==-(-y[24]+2*h*(y[25]^2-
2.5)+y[26]),
-1*u25+4*h*y[26]*u26+1*u27==-(-y[25]+2*h*(y[26]^2-
2.6)+y[27]),
-1*u26+4*h*y[27]*u27+1*u28==-(-y[26]+2*h*(y[27]^2-
2.7)+y[28]),
-1*u27+4*h*y[28]*u28+1*u29==-(-y[27]+2*h*(y[28]^2-
2.8)+y[29]),
-1*u28+4*h*y[29]*u29+1*u30==-(-y[28]+2*h*(y[29]^2-
2.9)+y[30]),
-1*u29+4*h*y[30]*u30+1*u31==-(-y[29]+2*h*(y[30]^2-
3.0)+y[31]),
-1*u30+4*h*y[31]*u31+1*u32==-(-y[30]+2*h*(y[31]^2-
3.1)+y[32]),
-1*u31+4*h*y[32]*u32+1*u33==-(-y[31]+2*h*(y[32]^2-
3.2)+y[33]),
-1*u32+4*h*y[33]*u33+1*u34==-(-y[32]+2*h*(y[33]^2-
3.3)+y[34]),
-1*u33+4*h*y[34]*u34+1*u35==-(-y[33]+2*h*(y[34]^2-
3.4)+y[35]),
-1*u34+4*h*y[35]*u35+1*u36==-(-y[34]+2*h*(y[35]^2-
3.5)+y[36]),
-1*u35+4*h*y[36]*u36+1*u37==-(-y[35]+2*h*(y[36]^2-
3.6)+y[37]),
-1*u36+4*h*y[37]*u37+1*u38==-(-y[36]+2*h*(y[37]^2-
3.7)+y[38]),
-1*u37+4*h*y[38]*u38+1*u39==-(-y[37]+2*h*(y[38]^2-
3.8)+y[39]),
-1*u38+4*h*y[39]*u39==-(-y[38]+2*h*(y[39]^2-
3.9)+y[40])},
{u1,u2,u3,u4,u5,u6,u7,u8,u9,u10,u11,u12,u13,u14,u15
,u16,u17,u18,u19,u20,
u21,u22,u23,u24,u25,u26,u27,u28,u29,u30,u31,u32,u33
,u34,u35,u36,u37,u38,u39}];

f[1]=N[EL1=Part[G,1];L1=u1/.EL1];
f[2]=N[EL2=Part[G,1];L2=u2/.EL2];
f[3]=N[EL3=Part[G,1];L3=u3/.EL3];
f[4]=N[EL4=Part[G,1];L4=u4/.EL4];
f[5]=N[EL5=Part[G,1];L5=u5/.EL5];
f[6]=N[EL6=Part[G,1];L6=u6/.EL6];
f[7]=N[EL7=Part[G,1];L7=u7/.EL7];
```

```
f[8]=N[EL8=Part[G,1];L8=u8/.EL8];
f[9]=N[EL9=Part[G,1];L9=u9/.EL9];
f[10]=N[EL10=Part[G,1];L10=u10/.EL10];
f[11]=N[EL11=Part[G,1];L11=u11/.EL11];
f[12]=N[EL12=Part[G,1];L12=u12/.EL12];
f[13]=N[EL13=Part[G,1];L13=u13/.EL13];
f[14]=N[EL14=Part[G,1];L14=u14/.EL14];
f[15]=N[EL15=Part[G,1];L15=u15/.EL15];
f[16]=N[EL16=Part[G,1];L16=u16/.EL16];
f[17]=N[EL17=Part[G,1];L17=u17/.EL17];
f[18]=N[EL18=Part[G,1];L18=u18/.EL18];
f[19]=N[EL19=Part[G,1];L19=u19/.EL19];
f[20]=N[EL20=Part[G,1];L20=u20/.EL20];
f[21]=N[EL21=Part[G,1];L21=u21/.EL21];
f[22]=N[EL22=Part[G,1];L22=u22/.EL22];
f[23]=N[EL23=Part[G,1];L23=u23/.EL23];
f[24]=N[EL24=Part[G,1];L24=u24/.EL24];
f[25]=N[EL25=Part[G,1];L25=u25/.EL25];
f[26]=N[EL26=Part[G,1];L26=u26/.EL26];
f[27]=N[EL27=Part[G,1];L27=u27/.EL27];
f[28]=N[EL28=Part[G,1];L28=u28/.EL28];
f[29]=N[EL29=Part[G,1];L29=u29/.EL29];
f[30]=N[EL30=Part[G,1];L30=u30/.EL30];
f[31]=N[EL31=Part[G,1];L31=u31/.EL31];
f[32]=N[EL32=Part[G,1];L32=u32/.EL32];
f[33]=N[EL33=Part[G,1];L33=u33/.EL33];
f[34]=N[EL34=Part[G,1];L34=u34/.EL34];
f[35]=N[EL35=Part[G,1];L35=u35/.EL35];
f[36]=N[EL36=Part[G,1];L36=u36/.EL36];
f[37]=N[EL37=Part[G,1];L37=u37/.EL37];
f[38]=N[EL38=Part[G,1];L38=u38/.EL38];
f[39]=N[EL39=Part[G,1];L39=u39/.EL39];

y[1]=f[1]+y[1];
y[2]=f[2]+y[2];
y[3]=f[3]+y[3];
y[4]=f[4]+y[4];y[5]=f[5]+y[5];
y[6]=f[6]+y[6];y[7]=f[7]+y[7];y[8]=f[8]+y[8];
y[9]=f[9]+y[9];y[10]=f[10]+y[10];
y[11]=f[11]+y[11];y[12]=f[12]+y[12];y[13]=f[13]+y[13];
y[14]=f[14]+y[14];
y[15]=f[15]+y[15];y[16]=f[16]+y[16];y[17]=f[17]+
y[17];y[18]=f[18]+y[18];
y[19]=f[19]+y[19];y[20]=f[20]+y[20];y[21]=f[21]+
y[21];y[22]=f[22]+y[22];
```

```
y[23]=f[23]+y[23];y[24]=f[24]+y[24];y[25]=f[25]+
y[25];y[26]=f[26]+y[26];
y[27]=f[27]+y[27];y[28]=f[28]+y[28];y[29]=f[29]+
y[29];y[30]=f[30]+y[30];
y[31]=f[31]+y[31];y[32]=f[32]+y[32];y[33]=f[33]+
y[33];y[34]=f[34]+y[34];
y[35]=f[35]+y[35];y[36]=f[36]+y[36];y[37]=f[37]+
y[37];y[38]=f[38]+y[38];
y[39]=f[39]+y[39];

x=0;
i=0;
Table[{i=i+1,x=x+h,y[i]},{i,0,38,1}];
TableForm[%,TableSpacing->{3,3},TableHeadings->
{None,{"i","x","y"}}]
x=0;
i=0;
p1=ListPlot[Table[{{i=i+1;x=x+h,y[i]}},
{i,0,38,1}],Frame->True,
FrameLabel->{"x","y"},FrameTicks->All,PlotStyle->
{Black},PlotRange->Automatic]

bbbbbbbbbbbbbbbbbbbbbbbbbbbbbbbbbbb
```

TABLE 2.2

The Results of the Program Number 2.2

i	x	Initial values of y_i's taken	y_i's after 4th iteration	Euler solution y_i's of Table 2.1
1	0.1	3.0	2.3415	2.1100
2	0.2	2.5	1.9235	1.6848
3	0.3	2.0	1.6416	1.4309
4	0.4	2.0	1.4445	1.2662
5	0.5	1.5	1.3043	1.1559
6	0.6	1.0	1.2043	1.0823
7	0.7	2.0	1.1342	1.0351
8	0.8	1.0	1.0870	1.0080
9	0.9	1.0	1.0579	0.9964
10	1.0	1.0	1.0432	0.9971
11	1.1	1.0	1.0402	1.0077

(continued)

TABLE 2.2 (Continued)

i	x	Initial values of y_i's taken	y_i's after 4th iteration	Euler solution y_i's of Table 2.1
12	1.2	1.0	1.0468	1.0261
13	1.3	1.0	1.0611	1.0508
14	1.4	1.0	1.0816	1.0804
15	1.5	1.0	1.1071	1.1137
16	1.6	1.0	1.1365	1.1497
17	1.7	1.0	1.1688	1.1875
18	1.8	1.0	1.2033	1.2265
19	1.9	1.0	1.2392	1.2661
20	2.0	1.0	1.2762	1.3058
21	2.1	0.5	1.3135	1.3453
22	2.2	0.5	1.3511	1.3843
23	2.3	2.5	1.3884	1.4227
24	2.4	0.5	1.4256	1.4603
25	2.5	0.5	1.4619	1.4970
26	2.6	2.5	1.4982	1.5329
27	2.7	0.5	1.5329	1.5679
28	2.8	2.5	1.5683	1.6021
29	2.9	0.5	1.6010	1.6354
30	3.0	0.5	1.6356	1.6680
31	3.1	0.5	1.6660	1.6998
32	3.2	0.5	1.7005	1.7308
33	3.3	2.5	1.7277	1.7613
34	3.4	2.5	1.7635	1.7911
35	3.5	3.5	1.7857	1.8203
36	3.6	3.5	1.8258	1.8489
37	3.7	4.0	1.8390	1.8771
38	3.8	3.5	1.8894	1.9047
39	3.9	4.0	1.8851	1.9319

Numerical solution to equation (2.1): $\frac{dy}{dx} + y^2 = x$ with the boundary values $(x, y) = (0, 3)$ and $(4, 1.9587)$ using Newton's iterative method. Comparison with Euler solution with initial value $(x, y) = (0, 3)$ shows rapid convergence to Euler solution.

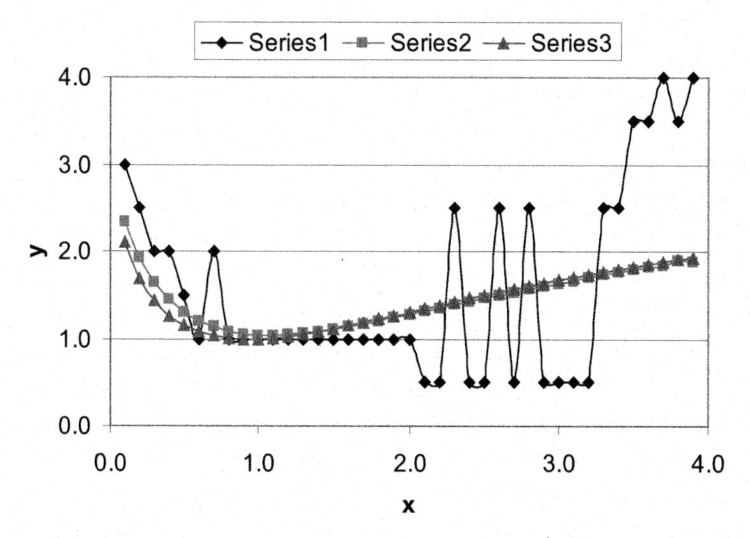

Figure 2.1 Numerical solution to equation (2.1): $\frac{dy}{dx} + y^2 = x$ with the boundary values $(x, y) = (0, 3)$ and $(4, 1.9587)$ using Newton's iterative method. Showing the results of program number 2.2. Series 1 shows the initial values of y_i's taken, series 2 shows the results of Newton's method y_i's after 4th iteration, and series 3 shows the Euler solution of Table 2.1. Rapid convergence to Euler solution is evident.

3

NUMERICAL SOLUTION OF BOUNDARY VALUE PROBLEM OF NON-LINEAR DIFFERENTIAL EQUATION

Example II

In this chapter, we have presented the numerical solution to boundary value problem of the 2nd non-linear differential equation in this book in comprehensive detail.

3.1 THE 2ND NON-LINEAR DIFFERENTIAL EQUATION IN THIS BOOK: EULER SOLUTION

In this chapter, we deal with the boundary value problem of the following non-linear differential equation:

$$\frac{dy}{dx} + y^2 = \text{Sin}(x) \tag{3.1}$$

We first turn to Euler solution of this differential equation. To this end, we re-write equation (3.1) as

$$\frac{dy}{dx} = \text{Sin}(x) - y^2 \tag{3.2}$$

which looks like equation (1.14): $\frac{dy}{dx} = f(x, y)$ which admits the iteration given by equation (1.20): $y_{n+1} = y_n + hf(x, y)$.

We have written program number 3.1 following Euler's method and solved equation (3.2) with initial values $(x, y) = (0, 0)$ with the numerical result shown in Table 3.1. From Table 3.1, we gather the boundary values $(x, y) = (0, 0)$ and $(4, 0.0850)$.

DOI: 10.1201/9781003204916-3

Program Number 3.1

```
h=0.1;
i=0;
x=0;
y=0;
Table[{i=i+1,x=x+h,y=y+h*(Sin[x]-y^2)},{x,0,4,h}];
TableForm[%,TableSpacing->{2,2},TableHeadings
->{None,{"i","x","y"}}]

i=0;
x=0;
y=0;
p1=ListPlot[Table[{i=i+1;x=x-
+h,y=y+h*(Sin[x]-y^2)},{x,0,4,h}],
Frame->True,FrameLabel->{"x","y"},FrameTicks
->All,PlotStyle->{Black}]
```

3.2 THE 2ND NON-LINEAR DIFFERENTIAL EQUATION IN THIS BOOK: SOLUTION BY NEWTON'S ITERATIVE METHOD

For now we assume that the Euler solution is unknown and we have the non-linear differential equation (3.1): $\frac{dy}{dx}+y^2=\mathrm{Sin}(x)$ with the boundary values $(x, y) = (0, 0)$ and $(4, 0.0850)$.

We proceed by replacing the derivative dy/dx with finite difference approximation given by equation (1.12): $y' = (y_{i+1} - y_{i-1})/(2h)$. This results in the following difference equation:

$$\frac{y_{i+1}-y_{i-1}}{2h}+y_i^2=\mathrm{Sin}(x_i)$$

or,

$$-y_{i-1}+2h(y_i^2-\mathrm{Sin}(x_i))+y_{i+1}=0 \tag{3.3}$$

We divide the interval $x = 0$ to 4 into 40 equal parts each being $h = 0.1$. We write equation (3.3) for $i = 1, 2, 3, \ldots, 39$ for which $x_i = 0.1, 0.2, 0.3, \ldots, 3.9$, respectively. We get the following system of non-linear algebraic equations:

$$f_1=-y_0+2h(y_1^2-\mathrm{Sin}(x_1))+y_2=0$$
$$f_2=-y_1+2h(y_2^2-\mathrm{Sin}(x_2))+y_3=0$$

$$f_3 = -y_2 + 2h(y_3^2 - \text{Sin}(x_3)) + y_4 = 0$$

$$\ldots$$

$$f_{39} = -y_{38} + 2h(y_{39}^2 - \text{Sin}(x_{39})) + y_{40} = 0$$

Here $y_0 = 0$ and $y_{40} = 0.0850$ are constants. As such, the corresponding system of linear equations following Newton's iterative method is as follows:

$$4hy_1u_1 + 1u_2 = -(-y_0 + 2h(y_1^2 - \text{Sin}(x_1)) + y_2)$$

$$-1u_1 + 4hy_2u_2 + 1u_3 = -(-y_1 + 2h(y_2^2 - \text{Sin}(x_2)) + y_3)$$

$$-1u_2 + 4hy_3u_3 + 1u_4 = -(-y_2 + 2h(y_3^2 - \text{Sin}(x_3)) + y_4)$$

$$\ldots$$

$$-1u_{38} + 4hy_{39}u_{39} = -(-y_{38} + 2h(y_{39}^2 - \text{Sin}(x_{39})) + y_{40})$$

To solve these equations which are linear in u_i's, we have taken initial set of values of y[i]'s as

```
y[0]=0;y[1]=0.3;y[2]=0.3;y[3]=0.3;y[4]=0.3;
y[5]=0.3;y[6]=0.3;y[7]=0.3;y[8]=0.3;y[9]=0.3;
y[10]=0.3;y[11]=0.5;y[12]=0.5;y[13]=0.5;y[14]=0.5;
y[15]=0.5;y[16]=0.5;y[17]=0.5;
y[18]=0.5;y[19]=0.7;y[20]=0.7;y[21]=0.7;y[22]=0.7;
y[23]=0.7;y[24]=0.7;y[25]=0.7;
y[26]=0.7;y[27]=0.3;y[28]=0.3;y[29]=0.3;y[30]=0.3;
y[31]=0.3;y[32]=0.3;y[33]=0.3;
y[34]=0.3;y[35]=0.3;y[36]=0.3;y[37]=0.3;y[38]=0.3;
y[39]=0.3;y[40]=0.0850;
```

in program number 3.2. In using program number 3.2, the content of the program between the lines marked as

```
aaaaaaaaaaaaaaaaaaaaaaaaaaaaaaa
```

and

```
bbbbbbbbbbbbbbbbbbbbbbbbbbbbbbbbbbbb
```

should be pasted p–1 times, after the line

```
bbbbbbbbbbbbbbbbbbbbbbbbbbbbbbbbbbbb
```

TABLE 3.1
Euler Solution of Equation (3.2): $\frac{dy}{dx}$ =Sin(x)− y^2 with Initial Condition $(x, y) = (0, 0)$

i	*x*	*y*
1	0.1	0.0100
2	0.2	0.0298
3	0.3	0.0593
4	0.4	0.0979
5	0.5	0.1449
6	0.6	0.1992
7	0.7	0.2597
8	0.8	0.3247
9	0.9	0.3925
10	1.0	0.4612
11	1.1	0.5291
12	1.2	0.5943
13	1.3	0.6553
14	1.4	0.7109
15	1.5	0.7601
16	1.6	0.8023
17	1.7	0.8371
18	1.8	0.8644
19	1.9	0.8843
20	2.0	0.8971
21	2.1	0.9029
22	2.2	0.9022
23	2.3	0.8954
24	2.4	0.8828
25	2.5	0.8647
26	2.6	0.8415
27	2.7	0.8134
28	2.8	0.7807
29	2.9	0.7437
30	3.0	0.7025
31	3.1	0.6573
32	3.2	0.6083

(continued)

i	x	y
33	3.3	0.5555
34	3.4	0.4991
35	3.5	0.4391
36	3.6	0.3756
37	3.7	0.3085
38	3.8	0.2378
39	3.9	0.1633
40	4.0	0.0850

where p (= 4 in this case) is the number of iterations required to produce convergence of the values of y[i]'s. **The results of the program are shown in Table 3.2 and Figure 3.1.**

Program Number 3.2

```
h=0.1;
y[0]=0;
y[40]=0.0850;
y[1]=0.3;y[2]=0.3;y[3]=0.3;y[4]=0.3;y[5]=0.3;
y[6]=0.3;y[7]=0.3;y[8]=0.3;y[9]=0.3;
y[10]=0.3;y[11]=0.5;y[12]=0.5;y[13]=0.5;y[14]=0.5;
y[15]=0.5;y[16]=0.5;y[17]=0.5;
y[18]=0.5;y[19]=0.7;y[20]=0.7;y[21]=0.7;y[22]=0.7;
y[23]=0.7;y[24]=0.7;y[25]=0.7;
y[26]=0.7;y[27]=0.3;y[28]=0.3;y[29]=0.3;y[30]=0.3;
y[31]=0.3;y[32]=0.3;y[33]=0.3;
y[34]=0.3;y[35]=0.3;y[36]=0.3;y[37]=0.3;
y[38]=0.3;y[39]=0.3;

x=0;
i=0;
Table[{i=i+1,x=x+h,y[i]},{i,0,38,1}];
TableForm[%,TableSpacing->{3,3},TableHeadings
->{None,{"i","x","y"}}]
x=0;
i=0;
p1=ListPlot[Table[{{i=i+1;x=x+h,
y[i]}},{i,0,38,1}],Frame->True,
FrameLabel->{"x","y"},FrameTicks->All,PlotStyle->
{Black},PlotRange->Automatic]

aaaaaaaaaaaaaaaaaaaaaaaaaaaaaaaa
```

```
G=NSolve[{
4*h*y[1]*u1+1*u2==-(-y[0]+2*h*(y[1]^2-
(Sin[0.1]))+y[2]),
-1*u1+4*h*y[2]*u2+1*u3==-(-y[1]+2*h*(y[2]^2-
(Sin[0.2]))+y[3]),
-1*u2+4*h*y[3]*u3+1*u4==-(-y[2]+2*h*(y[3]^2-
(Sin[0.3]))+y[4]),
-1*u3+4*h*y[4]*u4+1*u5==-(-y[3]+2*h*(y[4]^2-
(Sin[0.4]))+y[5]),
-1*u4+4*h*y[5]*u5+1*u6==-(-y[4]+2*h*(y[5]^2-
(Sin[0.5]))+y[6]),
-1*u5+4*h*y[6]*u6+1*u7==-(-y[5]+2*h*(y[6]^2-
(Sin[0.6]))+y[7]),
-1*u6+4*h*y[7]*u7+1*u8==-(-y[6]+2*h*(y[7]^2-
(Sin[0.7]))+y[8]),
-1*u7+4*h*y[8]*u8+1*u9==-(-y[7]+2*h*(y[8]^2-
(Sin[0.8]))+y[9]),
-1*u8+4*h*y[9]*u9+1*u10==-(-y[8]+2*h*(y[9]^2-
(Sin[0.9]))+y[10]),
-1*u9+4*h*y[10]*u10+1*u11==-(-y[9]+2*h*(y[10]^2-
(Sin[1.0]))+y[11]),
-1*u10+4*h*y[11]*u11+1*u12==-(-y[10]+2*h*(y[11]^2-
(Sin[1.1]))+y[12]),
-1*u11+4*h*y[12]*u12+1*u13==-(-y[11]+2*h*(y[12]^2-
(Sin[1.2]))+y[13]),
-1*u12+4*h*y[13]*u13+1*u14==-(-y[12]+2*h*(y[13]^2-
(Sin[1.3]))+y[14]),
-1*u13+4*h*y[14]*u14+1*u15==-(-y[13]+2*h*(y[14]^2-
(Sin[1.4]))+y[15]),
-1*u14+4*h*y[15]*u15+1*u16==-(-y[14]+2*h*(y[15]^2-
(Sin[1.5]))+y[16]),
-1*u15+4*h*y[16]*u16+1*u17==-(-y[15]+2*h*(y[16]^2-
(Sin[1.6]))+y[17]),
-1*u16+4*h*y[17]*u17+1*u18==-(-y[16]+2*h*(y[17]^2-
(Sin[1.7]))+y[18]),
-1*u17+4*h*y[18]*u18+1*u19==-(-y[17]+2*h*(y[18]^2-
(Sin[1.8]))+y[19]),
-1*u18+4*h*y[19]*u19+1*u20==-(-y[18]+2*h*(y[19]^2-
(Sin[1.9]))+y[20]),
-1*u19+4*h*y[20]*u20+1*u21==-(-y[19]+2*h*(y[20]^2-
(Sin[2.0]))+y[21]),
-1*u20+4*h*y[21]*u21+1*u22==-(-y[20]+2*h*(y[21]^2-
(Sin[2.1]))+y[22]),
-1*u21+4*h*y[22]*u22+1*u23==-(-y[21]+2*h*(y[22]^2-
(Sin[2.2]))+y[23]),
-1*u22+4*h*y[23]*u23+1*u24==-(-y[22]+2*h*(y[23]^2-
(Sin[2.3]))+y[24]),
```

```
-1*u23+4*h*y[24]*u24+1*u25==-(-y[23]+2*h*(y[24]^2-
(Sin[2.4]))+y[25]),
-1*u24+4*h*y[25]*u25+1*u26==-(-y[24]+2*h*(y[25]^2-
(Sin[2.5]))+y[26]),
-1*u25+4*h*y[26]*u26+1*u27==-(-y[25]+2*h*(y[26]^2-
(Sin[2.6]))+y[27]),
-1*u26+4*h*y[27]*u27+1*u28==-(-y[26]+2*h*(y[27]^2-
(Sin[2.7]))+y[28]),
-1*u27+4*h*y[28]*u28+1*u29==-(-y[27]+2*h*(y[28]^2-
(Sin[2.8]))+y[29]),
-1*u28+4*h*y[29]*u29+1*u30==-(-y[28]+2*h*(y[29]^2-
(Sin[2.9]))+y[30]),
-1*u29+4*h*y[30]*u30+1*u31==-(-y[29]+2*h*(y[30]^2-
(Sin[3.0]))+y[31]),
-1*u30+4*h*y[31]*u31+1*u32==-(-y[30]+2*h*(y[31]^2-
(Sin[3.1]))+y[32]),
-1*u31+4*h*y[32]*u32+1*u33==-(-y[31]+2*h*(y[32]^2-
(Sin[3.2]))+y[33]),
-1*u32+4*h*y[33]*u33+1*u34==-(-y[32]+2*h*(y[33]^2-
(Sin[3.3]))+y[34]),
-1*u33+4*h*y[34]*u34+1*u35==-(-y[33]+2*h*(y[34]^2-
(Sin[3.4]))+y[35]),
-1*u34+4*h*y[35]*u35+1*u36==-(-y[34]+2*h*(y[35]^2-
(Sin[3.5]))+y[36]),
-1*u35+4*h*y[36]*u36+1*u37==-(-y[35]+2*h*(y[36]^2-
(Sin[3.6]))+y[37]),
-1*u36+4*h*y[37]*u37+1*u38==-(-y[36]+2*h*(y[37]^2-
(Sin[3.7]))+y[38]),
-1*u37+4*h*y[38]*u38+1*u39==-(-y[37]+2*h*(y[38]^2-
(Sin[3.8]))+y[39]),
-1*u38+4*h*y[39]*u39==-(-y[38]+2*h*(y[39]^2-
(Sin[3.9]))+y[40])},
{u1,u2,u3,u4,u5,u6,u7,u8,u9,u10,u11,u12,u13,u14,u15
,u16,u17,u18,u19,u20,
u21,u22,u23,u24,u25,u26,u27,u28,u29,u30,u31,u32,u33
,u34,u35,u36,u37,u38,u39}];

f[1]=N[EL1=Part[G,1];L1=u1/.EL1];
f[2]=N[EL2=Part[G,1];L2=u2/.EL2];
f[3]=N[EL3=Part[G,1];L3=u3/.EL3];
f[4]=N[EL4=Part[G,1];L4=u4/.EL4];
f[5]=N[EL5=Part[G,1];L5=u5/.EL5];
f[6]=N[EL6=Part[G,1];L6=u6/.EL6];
f[7]=N[EL7=Part[G,1];L7=u7/.EL7];
f[8]=N[EL8=Part[G,1];L8=u8/.EL8];
f[9]=N[EL9=Part[G,1];L9=u9/.EL9];
```

```
f[10]=N[EL10=Part[G,1];L10=u10/.EL10];
f[11]=N[EL11=Part[G,1];L11=u11/.EL11];
f[12]=N[EL12=Part[G,1];L12=u12/.EL12];
f[13]=N[EL13=Part[G,1];L13=u13/.EL13];
f[14]=N[EL14=Part[G,1];L14=u14/.EL14];
f[15]=N[EL15=Part[G,1];L15=u15/.EL15];
f[16]=N[EL16=Part[G,1];L16=u16/.EL16];
f[17]=N[EL17=Part[G,1];L17=u17/.EL17];
f[18]=N[EL18=Part[G,1];L18=u18/.EL18];
f[19]=N[EL19=Part[G,1];L19=u19/.EL19];
f[20]=N[EL20=Part[G,1];L20=u20/.EL20];
f[21]=N[EL21=Part[G,1];L21=u21/.EL21];
f[22]=N[EL22=Part[G,1];L22=u22/.EL22];
f[23]=N[EL23=Part[G,1];L23=u23/.EL23];
f[24]=N[EL24=Part[G,1];L24=u24/.EL24];
f[25]=N[EL25=Part[G,1];L25=u25/.EL25];
f[26]=N[EL26=Part[G,1];L26=u26/.EL26];
f[27]=N[EL27=Part[G,1];L27=u27/.EL27];
f[28]=N[EL28=Part[G,1];L28=u28/.EL28];
f[29]=N[EL29=Part[G,1];L29=u29/.EL29];
f[30]=N[EL30=Part[G,1];L30=u30/.EL30];
f[31]=N[EL31=Part[G,1];L31=u31/.EL31];
f[32]=N[EL32=Part[G,1];L32=u32/.EL32];
f[33]=N[EL33=Part[G,1];L33=u33/.EL33];
f[34]=N[EL34=Part[G,1];L34=u34/.EL34];
f[35]=N[EL35=Part[G,1];L35=u35/.EL35];
f[36]=N[EL36=Part[G,1];L36=u36/.EL36];
f[37]=N[EL37=Part[G,1];L37=u37/.EL37];
f[38]=N[EL38=Part[G,1];L38=u38/.EL38];
f[39]=N[EL39=Part[G,1];L39=u39/.EL39];

y[1]=f[1]+y[1];
y[2]=f[2]+y[2];
y[3]=f[3]+y[3];
y[4]=f[4]+y[4];y[5]=f[5]+y[5];
y[6]=f[6]+y[6];y[7]=f[7]+y[7];y[8]=f[8]+y[8];y[9]=
f[9]+y[9];y[10]=f[10]+y[10];
y[11]=f[11]+y[11];y[12]=f[12]+y[12];y[13]=f[13]+
y[13];y[14]=f[14]+y[14];
y[15]=f[15]+y[15];y[16]=f[16]+y[16];y[17]=f[17]+
y[17];y[18]=f[18]+y[18];
y[19]=f[19]+y[19];y[20]=f[20]+y[20];y[21]=f[21]+
y[21];y[22]=f[22]+y[22];
y[23]=f[23]+y[23];y[24]=f[24]+y[24];y[25]=f[25]+
y[25];y[26]=f[26]+y[26];
y[27]=f[27]+y[27];y[28]=f[28]+y[28];y[29]=f[29]+
y[29];y[30]=f[30]+y[30];
```

```
y[31]=f[31]+y[31];y[32]=f[32]+y[32];y[33]=f[33]+
y[33];y[34]=f[34]+y[34];
y[35]=f[35]+y[35];y[36]=f[36]+y[36];y[37]=f[37]+
y[37];y[38]=f[38]+y[38];
y[39]=f[39]+y[39];

x=0;
i=0;
Table[{i=i+1,x=x+h,y[i]},{i,0,38,1}];
TableForm[%,TableSpacing->{3,3},TableHeadings
->{None,{"i","x","y"}}]

x=0;
i=0;
p1=ListPlot[Table[{{i=i+1;x=x+h,y[i]}},{i,0,38,1}],
Frame->True,
FrameLabel->{"x","y"},FrameTicks->All,PlotStyle->
{Black},PlotRange->Automatic]
bbbbbbbbbbbbbbbbbbbbbbbbbbbbbbbbbbbbb
```

TABLE 3.2

The Results of the Program Number 3.2

i	x	Initial values of y_i's taken	y_i's after 4th iteration	Euler solution of Table 3.1
1	0.1	0.3	0.0066	0.0100
2	0.2	0.3	0.0200	0.0298
3	0.3	0.3	0.0462	0.0593
4	0.4	0.3	0.0786	0.0979
5	0.5	0.3	0.1229	0.1449
6	0.6	0.3	0.1715	0.1992
7	0.7	0.3	0.2299	0.2597
8	0.8	0.3	0.2898	0.3247
9	0.9	0.3	0.3566	0.3925
10	1.0	0.3	0.4210	0.4612
11	1.1	0.5	0.4894	0.5291
12	1.2	0.5	0.5513	0.5943
13	1.3	0.5	0.6150	0.6553
14	1.4	0.5	0.6684	0.7109

(continued)

TABLE 3.2 (Continued)

i	x	Initial values of y_i's taken	y_i's after 4th iteration	Euler solution of Table 3.1
15	1.5	0.5	0.7228	0.7601
16	1.6	0.5	0.7634	0.8023
17	1.7	0.5	0.8061	0.8371
18	1.8	0.5	0.8318	0.8644
19	1.9	0.7	0.8625	0.8843
20	2.0	0.7	0.8722	0.8971
21	2.1	0.7	0.8922	0.9029
22	2.2	0.7	0.8857	0.9022
23	2.3	0.7	0.8971	0.8954
24	2.4	0.7	0.8739	0.8828
25	2.5	0.7	0.8794	0.8647
26	2.6	0.7	0.8389	0.8415
27	2.7	0.3	0.8418	0.8134
28	2.8	0.3	0.7826	0.7807
29	2.9	0.3	0.7863	0.7437
30	3.0	0.3	0.7068	0.7025
31	3.1	0.3	0.7146	0.6573
32	3.2	0.3	0.6130	0.6083
33	3.3	0.3	0.6277	0.5555
34	3.4	0.3	0.5027	0.4991
35	3.5	0.3	0.5261	0.4391
36	3.6	0.3	0.3772	0.3756
37	3.7	0.3	0.4091	0.3085
38	3.8	0.3	0.2377	0.2378
39	3.9	0.3	0.2755	0.1633

Numerical solution to equation (3.1): $\frac{dy}{dx} + y^2 = Sin(x)$ with the boundary values $(x, y) = (0, 0)$ and $(4, 0.0850)$ using Newton's iterative method. Comparison with Euler solution with initial value $(x, y) = (0, 0)$ shows rapid convergence to Euler solution.

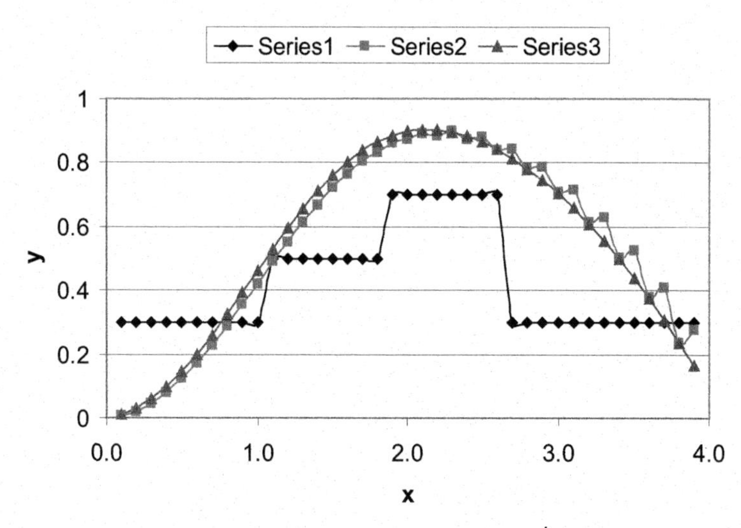

Figure 3.1 Numerical solution to equation (3.1): $\frac{dy}{dx} + y^2 = Sin(x)$ with the boundary values $(x, y) = (0, 0)$ and $(4, 0.0850)$ using Newton's iterative method. Showing the results of program number 3.2. Series 1 shows the initial values of y_i's taken, series 2 shows the results of Newton's method y_i's after 4th iteration, and series 3 shows the Euler solution of Table 3.1. Rapid convergence to Euler solution is evident. (The small oscillations in series 2 do not disappear even if we raise number of iterations.)

4

NUMERICAL SOLUTION OF BOUNDARY VALUE PROBLEM OF NON-LINEAR DIFFERENTIAL EQUATION

Example III

In this chapter, we have presented the numerical solution to boundary value problem of the 3rd non-linear differential equation in this book in comprehensive detail.

4.1 THE 3RD NON-LINEAR DIFFERENTIAL EQUATION IN THIS BOOK: EULER SOLUTION

In this chapter, we deal with the boundary value problem of the following non-linear differential equation:

$$\frac{dy}{dx} + y^2 = \mathrm{Sin}^2(x) \tag{4.1}$$

We first turn to Euler solution of this differential equation. To this end, we re-write equation (4.1) as

$$\frac{dy}{dx} = \mathrm{Sin}^2(x) - y^2 \tag{4.2}$$

which looks like equation (1.14): $\frac{dy}{dx} = f(x, y)$ which admits the iteration given by equation (1.20): $y_{n+1} = y_n + hf(x, y)$.

We have written program number 4.1 following Euler's method and solved equation (4.2) with initial values $(x, y) = (0, 0)$ with the

DOI: 10.1201/9781003204916-4

numerical result shown in Table 4.1. From Table 4.1, we gather the boundary values $(x, y) = (0, 0)$ and $(8, 0.9033)$.

Program Number 4.1

```
h=0.2;
i=0;
x=0;
y=0;
Table[{i=i+1,x=x+h,y=y+h*((Sin[x])^2-y^2)},
{x,0,8,h}];
TableForm[%,TableSpacing->{2,2},TableHeadings->
{None,{"i","x","y"}}]

i=0;
x=0;
y=0;
p1=ListPlot[Table[{i=i+1;x=x+h,y=y+h*((Sin[x])
^2-y^2)},{x,0,8,h}],
Frame->True,FrameLabel->{"x","y"},FrameTicks->
All,PlotStyle->{Black}]
```

TABLE 4.1

Euler Solution of Equation (4.2): $\frac{dy}{dx} = Sin^2(x) - y^2$ with Initial Condition $(x, y) = (0, 0)$

i	*x*	*Y*
1	0.2	0.0079
2	0.4	0.0382
3	0.6	0.1017
4	0.8	0.2025
5	1.0	0.3359
6	1.2	0.4871
7	1.4	0.6339
8	1.6	0.7533
9	1.8	0.8295
10	2.0	0.8573
11	2.2	0.8410
12	2.4	0.7908
13	2.6	0.7189

(continued)

i	*X*	*Y*
14	2.8	0.6380
15	3.0	0.5605
16	3.2	0.4984
17	3.4	0.4618
18	3.6	0.4583
19	3.8	0.4912
20	4.0	0.5575
21	4.2	0.6472
22	4.4	0.7446
23	4.6	0.8312
24	4.8	0.8915
25	5.0	0.9164
26	5.2	0.9046
27	5.4	0.8603
28	5.6	0.7920
29	5.8	0.7097
30	6.0	0.6246
31	6.2	0.5480
32	6.4	0.4906
33	6.6	0.4619
34	6.8	0.4681
35	7.0	0.5106
36	7.2	0.5844
37	7.4	0.6776
38	7.6	0.7732
39	7.8	0.8530
40	8.0	0.9033

4.2 THE 3RD NON-LINEAR DIFFERENTIAL EQUATION IN THIS BOOK: SOLUTION BY NEWTON'S ITERATIVE METHOD

For now we assume that the Euler solution is unknown and we have the non-linear differential equation (4.1): $\frac{dy}{dx} + y^2 = \mathrm{Sin}^2(x)$ with the boundary values $(x, y) = (0, 0)$ and $(8, 0.9033)$.

We proceed by replacing the derivative dy/dx with finite difference approximation given by equation (1.12): $y' = (y_{i+1} - y_{i-1})/(2h)$. This results in the following difference equation:

$$\frac{y_{i+1} - y_{i-1}}{2h} + y_i^2 = \mathrm{Sin}^2(x_i)$$

or,

$$-y_{i-1} + 2h(y_i^2 - \mathrm{Sin}^2(x_i)) + y_{i+1} = 0 \qquad (4.3)$$

We divide the interval $x = 0$ to 8 into 40 equal parts each being $h = 0.2$. We write equation (4.3) for $i = 1, 2, 3, \ldots, 39$ for which $x_i = 0.2, 0.4, 0.6, \ldots, 7.8$, respectively. We get the following system of non-linear algebraic equations:

$$\begin{aligned}
f_1 &= -y_0 + 2h(y_1^2 - \mathrm{Sin}^2(x_1)) + y_2 = 0 \\
f_2 &= -y_1 + 2h(y_2^2 - \mathrm{Sin}^2(x_2)) + y_3 = 0 \\
f_3 &= -y_2 + 2h(y_3^2 - \mathrm{Sin}^2(x_3)) + y_4 = 0 \\
&\ldots \\
f_{39} &= -y_{38} + 2h(y_{39}^2 - \mathrm{Sin}^2(x_{39})) + y_{40} = 0
\end{aligned}$$

Here $y_0 = 0$ and $y_{40} = 0.9033$ are constants. As such, the corresponding system of linear equations following Newton's iterative method is as follows:

$$\begin{aligned}
4hy_1u_1 + 1u_2 &= -(-y_0 + 2h(y_1^2 - \mathrm{Sin}^2(x_1)) + y_2) \\
-1u_1 + 4hy_2u_2 + 1u_3 &= -(-y_1 + 2h(y_2^2 - \mathrm{Sin}^2(x_2)) + y_3) \\
-1u_2 + 4hy_3u_3 + 1u_4 &= -(-y_2 + 2h(y_3^2 - \mathrm{Sin}^2(x_3)) + y_4) \\
&\ldots \\
-1u_{38} + 4hy_{39}u_{39} &= -(-y_{38} + 2h(y_{39}^2 - \mathrm{Sin}^2(x_{39})) + y_{40})
\end{aligned}$$

To solve these equations which are linear in u_i's, we have taken initial set of values of y[i]'s as

```
y[0]=0;y[1]=0.5;y[2]=0.5;y[3]=0.5;y[4]=0.5;
y[5]=0.5;y[6]=0.5;y[7]=0.5;y[8]=0.5;y[9]=0.5;
y[10]=0.5;y[11]=0.7;y[12]=0.7;y[13]=0.7;y[14]=0.7;
y[15]=0.7;y[16]=0.7;y[17]=0.7;
y[18]=0.7;y[19]=0.7;y[20]=0.7;y[21]=0.8;y[22]=0.8;
y[23]=0.8;y[24]=0.8;y[25]=0.8;
y[26]=0.8;y[27]=0.8;y[28]=0.8;y[29]=0.8;y[30]=0.8;
y[31]=0.9;y[32]=0.9;y[33]=0.9;
y[34]=0.9;y[35]=0.9;y[36]=0.9;y[37]=0.9;y[38]=0.9;
```

```
y[39]=0.9;y[40]=0.9033;
```

in program number 4.2. In using program number 4.2, the content of the program between the lines marked as

```
aaaaaaaaaaaaaaaaaaaaaaaaaaaaaaaaaaaaaaaa
```

and

```
bbbbbbbbbbbbbbbbbbbbbbbbbbbbbbbbbbbbbb
```

should be pasted $p-1$ times, after the line

```
bbbbbbbbbbbbbbbbbbbbbbbbbbbbbbbbbbbbbb
```

where p (= 3 in this case) is the number of iterations required to produce convergence of the values of y[i]'s. **The results of the program are shown in Table 4.2 and Figure 4.1.**

Program Number 4.2

```
h=0.2;
y[0]=0;
y[40]=0.9033;

y[1]=0.5;y[2]=0.5;y[3]=0.5;y[4]=0.5;y[5]=0.5;
y[6]=0.5;y[7]=0.5;y[8]=0.5;y[9]=0.5;
y[10]=0.5;y[11]=0.7;y[12]=0.7;y[13]=0.7;y[14]=0.7;
y[15]=0.7;y[16]=0.7;y[17]=0.7;
y[18]=0.7;y[19]=0.7;y[20]=0.7;y[21]=0.8;y[22]=0.8;
y[23]=0.8;y[24]=0.8;y[25]=0.8;
y[26]=0.8;y[27]=0.8;y[28]=0.8;y[29]=0.8;y[30]=0.8;
y[31]=0.9;y[32]=0.9;y[33]=0.9;
y[34]=0.9;y[35]=0.9;y[36]=0.9;y[37]=0.9;
y[38]=0.9;y[39]=0.9;

x=0;
i=0;
Table[{i=i+1,x=x+h,y[i]},{i,0,38,1}];
TableForm[%,TableSpacing->{3,3},TableHead-
ings->{None,{"i","x","y"}}]

x=0;
i=0;
p1=ListPlot[Table[{{i=i+1;x=x+h,
y[i]}},{i,0,38,1}],Frame->True,
FrameLabel->{"x","y"},FrameTicks->All,PlotStyle->
{Black},PlotRange->Automatic]
```

```
aaaaaaaaaaaaaaaaaaaaaaaaaaaaaaaaaaa
G=NSolve[{
4*h*y[1]*u1+1*u2==-(-y[0]+2*h*(y[1]^2-
(Sin[0.2])^2)+y[2]),
-1*u1+4*h*y[2]*u2+1*u3==-(-y[1]+2*h*(y[2]^2-
(Sin[0.4])^2)+y[3]),
-1*u2+4*h*y[3]*u3+1*u4==-(-y[2]+2*h*(y[3]^2-
(Sin[0.6])^2)+y[4]),
-1*u3+4*h*y[4]*u4+1*u5==-(-y[3]+2*h*(y[4]^2-
(Sin[0.8])^2)+y[5]),
-1*u4+4*h*y[5]*u5+1*u6==-(-y[4]+2*h*(y[5]^2-
(Sin[1.0])^2)+y[6]),
-1*u5+4*h*y[6]*u6+1*u7==-(-y[5]+2*h*(y[6]^2-
(Sin[1.2])^2)+y[7]),
-1*u6+4*h*y[7]*u7+1*u8==-(-y[6]+2*h*(y[7]^2-
(Sin[1.4])^2)+y[8]),
-1*u7+4*h*y[8]*u8+1*u9==-(-y[7]+2*h*(y[8]^2-
(Sin[1.6])^2)+y[9]),
-1*u8+4*h*y[9]*u9+1*u10==-(-y[8]+2*h*(y[9]^2-
(Sin[1.8])^2)+y[10]),
-1*u9+4*h*y[10]*u10+1*u11==-(-y[9]+2*h*(y[10]^2-
(Sin[2.0])^2)+y[11]),
-1*u10+4*h*y[11]*u11+1*u12==-(-y[10]+2*h*(y[11]^2-
(Sin[2.2])^2)+y[12]),
-1*u11+4*h*y[12]*u12+1*u13==-(-y[11]+2*h*(y[12]^2-
(Sin[2.4])^2)+y[13]),
-1*u12+4*h*y[13]*u13+1*u14==-(-y[12]+2*h*(y[13]^2-
(Sin[2.6])^2)+y[14]),
-1*u13+4*h*y[14]*u14+1*u15==-(-y[13]+2*h*(y[14]^2-
(Sin[2.8])^2)+y[15]),
-1*u14+4*h*y[15]*u15+1*u16==-(-y[14]+2*h*(y[15]^2-
(Sin[3.0])^2)+y[16]),
-1*u15+4*h*y[16]*u16+1*u17==-(-y[15]+2*h*(y[16]^2-
(Sin[3.2])^2)+y[17]),
-1*u16+4*h*y[17]*u17+1*u18==-(-y[16]+2*h*(y[17]^2-
(Sin[3.4])^2)+y[18]),
-1*u17+4*h*y[18]*u18+1*u19==-(-y[17]+2*h*(y[18]^2-
(Sin[3.6])^2)+y[19]),
-1*u18+4*h*y[19]*u19+1*u20==-(-y[18]+2*h*(y[19]^2-
(Sin[3.8])^2)+y[20]),
-1*u19+4*h*y[20]*u20+1*u21==-(-y[19]+2*h*(y[20]^2-
(Sin[4.0])^2)+y[21]),
-1*u20+4*h*y[21]*u21+1*u22==-(-y[20]+2*h*(y[21]^2-
(Sin[4.2])^2)+y[22]),
```

```
-1*u21+4*h*y[22]*u22+1*u23==-(-y[21]+2*h*(y[22]^2-
(Sin[4.4])^2)+y[23]),
-1*u22+4*h*y[23]*u23+1*u24==-(-y[22]+2*h*(y[23]^2-
(Sin[4.6])^2)+y[24]),
-1*u23+4*h*y[24]*u24+1*u25==-(-y[23]+2*h*(y[24]^2-
(Sin[4.8])^2)+y[25]),
-1*u24+4*h*y[25]*u25+1*u26==-(-y[24]+2*h*(y[25]^2-
(Sin[5.0])^2)+y[26]),
-1*u25+4*h*y[26]*u26+1*u27==-(-y[25]+2*h*(y[26]^2-
(Sin[5.2])^2)+y[27]),
-1*u26+4*h*y[27]*u27+1*u28==-(-y[26]+2*h*(y[27]^2-
(Sin[5.4])^2)+y[28]),
-1*u27+4*h*y[28]*u28+1*u29==-(-y[27]+2*h*(y[28]^2-
(Sin[5.6])^2)+y[29]),
-1*u28+4*h*y[29]*u29+1*u30==-(-y[28]+2*h*(y[29]^2-
(Sin[5.8])^2)+y[30]),
-1*u29+4*h*y[30]*u30+1*u31==-(-y[29]+2*h*(y[30]^2-
(Sin[6.0])^2)+y[31]),
-1*u30+4*h*y[31]*u31+1*u32==-(-y[30]+2*h*(y[31]^2-
(Sin[6.2])^2)+y[32]),
-1*u31+4*h*y[32]*u32+1*u33==-(-y[31]+2*h*(y[32]^2-
(Sin[6.4])^2)+y[33]),
-1*u32+4*h*y[33]*u33+1*u34==-(-y[32]+2*h*(y[33]^2-
(Sin[6.6])^2)+y[34]),
-1*u33+4*h*y[34]*u34+1*u35==-(-y[33]+2*h*(y[34]^2-
(Sin[6.8])^2)+y[35]),
-1*u34+4*h*y[35]*u35+1*u36==-(-y[34]+2*h*(y[35]^2-
(Sin[7.0])^2)+y[36]),
-1*u35+4*h*y[36]*u36+1*u37==-(-y[35]+2*h*(y[36]^2-
(Sin[7.2])^2)+y[37]),
-1*u36+4*h*y[37]*u37+1*u38==-(-y[36]+2*h*(y[37]^2-
(Sin[7.4])^2)+y[38]),
-1*u37+4*h*y[38]*u38+1*u39==-(-y[37]+2*h*(y[38]^2-
(Sin[7.6])^2)+y[39]),
-1*u38+4*h*y[39]*u39==-(-y[38]+2*h*(y[39]^2-
(Sin[7.8])^2)+y[40])},
{u1,u2,u3,u4,u5,u6,u7,u8,u9,u10,u11,u12,u13,u14,u15
,u16,u17,u18,u19,u20,
u21,u22,u23,u24,u25,u26,u27,u28,u29,u30,u31,u32,u33
,u34,u35,u36,u37,u38,u39}];

f[1]=N[EL1=Part[G,1];L1=u1/.EL1];
f[2]=N[EL2=Part[G,1];L2=u2/.EL2];
f[3]=N[EL3=Part[G,1];L3=u3/.EL3];
```

```
f[4]=N[EL4=Part[G,1];L4=u4/.EL4];
f[5]=N[EL5=Part[G,1];L5=u5/.EL5];
f[6]=N[EL6=Part[G,1];L6=u6/.EL6];
f[7]=N[EL7=Part[G,1];L7=u7/.EL7];
f[8]=N[EL8=Part[G,1];L8=u8/.EL8];
f[9]=N[EL9=Part[G,1];L9=u9/.EL9];
f[10]=N[EL10=Part[G,1];L10=u10/.EL10];
f[11]=N[EL11=Part[G,1];L11=u11/.EL11];
f[12]=N[EL12=Part[G,1];L12=u12/.EL12];
f[13]=N[EL13=Part[G,1];L13=u13/.EL13];
f[14]=N[EL14=Part[G,1];L14=u14/.EL14];
f[15]=N[EL15=Part[G,1];L15=u15/.EL15];
f[16]=N[EL16=Part[G,1];L16=u16/.EL16];
f[17]=N[EL17=Part[G,1];L17=u17/.EL17];
f[18]=N[EL18=Part[G,1];L18=u18/.EL18];
f[19]=N[EL19=Part[G,1];L19=u19/.EL19];
f[20]=N[EL20=Part[G,1];L20=u20/.EL20];
f[21]=N[EL21=Part[G,1];L21=u21/.EL21];
f[22]=N[EL22=Part[G,1];L22=u22/.EL22];
f[23]=N[EL23=Part[G,1];L23=u23/.EL23];
f[24]=N[EL24=Part[G,1];L24=u24/.EL24];
f[25]=N[EL25=Part[G,1];L25=u25/.EL25];
f[26]=N[EL26=Part[G,1];L26=u26/.EL26];
f[27]=N[EL27=Part[G,1];L27=u27/.EL27];
f[28]=N[EL28=Part[G,1];L28=u28/.EL28];
f[29]=N[EL29=Part[G,1];L29=u29/.EL29];
f[30]=N[EL30=Part[G,1];L30=u30/.EL30];
f[31]=N[EL31=Part[G,1];L31=u31/.EL31];
f[32]=N[EL32=Part[G,1];L32=u32/.EL32];
f[33]=N[EL33=Part[G,1];L33=u33/.EL33];
f[34]=N[EL34=Part[G,1];L34=u34/.EL34];
f[35]=N[EL35=Part[G,1];L35=u35/.EL35];
f[36]=N[EL36=Part[G,1];L36=u36/.EL36];
f[37]=N[EL37=Part[G,1];L37=u37/.EL37];
f[38]=N[EL38=Part[G,1];L38=u38/.EL38];
f[39]=N[EL39=Part[G,1];L39=u39/.EL39];

y[1]=f[1]+y[1];
y[2]=f[2]+y[2];
y[3]=f[3]+y[3];
y[4]=f[4]+y[4];y[5]=f[5]+y[5];
y[6]=f[6]+y[6];y[7]=f[7]+y[7];y[8]=f[8]+y[8];
y[9]=f[9]+y[9];y[10]=f[10]+y[10];
y[11]=f[11]+y[11];y[12]=f[12]+y[12];y[13]=f[13]+
y[13];y[14]=f[14]+y[14];
```

```
y[15]=f[15]+y[15];y[16]=f[16]+y[16];y[17]=f[17]+
y[17];y[18]=f[18]+y[18];
y[19]=f[19]+y[19];y[20]=f[20]+y[20];y[21]=f[21]+
y[21];y[22]=f[22]+y[22];
y[23]=f[23]+y[23];y[24]=f[24]+y[24];y[25]=f[25]+
y[25];y[26]=f[26]+y[26];
y[27]=f[27]+y[27];y[28]=f[28]+y[28];y[29]=f[29]+
y[29];y[30]=f[30]+y[30];
y[31]=f[31]+y[31];y[32]=f[32]+y[32];y[33]=f[33]+
y[33];y[34]=f[34]+y[34];
y[35]=f[35]+y[35];y[36]=f[36]+y[36];y[37]=f[37]+
y[37];y[38]=f[38]+y[38];
y[39]=f[39]+y[39];
x=0;
i=0;
Table[{i=i+1,x=x+h,y[i]},{i,0,38,1}];
TableForm[%,TableSpacing->{3,3},TableHeadings
->{None,{"i","x","y"}}]
x=0;
i=0;
p1=ListPlot[Table[{{i=i+1;x=x+h,
y[i]}},{i,0,38,1}],Frame->True,
FrameLabel->{"x","y"},FrameTicks->All,PlotStyle->
{Black},PlotRange->Automatic]
bbbbbbbbbbbbbbbbbbbbbbbbbbbbbbbbbbbbb
```

TABLE 4.2

The Results of the Program Number 4.2

i	x	Initial values of y_i's taken	y_i's after 3rd iteration	Euler solution of Table 4.1
1	0.2	0.5	0.0000	0.0079
2	0.4	0.5	0.0158	0.0382
3	0.6	0.5	0.0605	0.1017
4	0.8	0.5	0.1419	0.2025
5	1.0	0.5	0.2583	0.3359
6	1.2	0.5	0.3984	0.4871
7	1.4	0.5	0.5424	0.6339
8	1.6	0.5	0.6692	0.7533
9	1.8	0.5	0.7629	0.8295

(continued)

TABLE 4.2 (Continued)

i	x	Initial values of y_i's taken	y_i's after 3rd iteration	Euler solution of Table 4.1
10	2.0	0.5	0.8158	0.8573
11	2.2	0.7	0.8275	0.8410
12	2.4	0.7	0.8034	0.7908
13	2.6	0.7	0.7518	0.7189
14	2.8	0.7	0.6836	0.6380
15	3.0	0.7	0.6098	0.5605
16	3.2	0.7	0.5428	0.4984
17	3.4	0.7	0.4933	0.4618
18	3.6	0.7	0.4716	0.4583
19	3.8	0.7	0.4827	0.4912
20	4.0	0.7	0.5281	0.5575
21	4.2	0.8	0.6002	0.6472
22	4.4	0.8	0.6879	0.7446
23	4.6	0.8	0.7731	0.8312
24	4.8	0.8	0.8438	0.8915
25	5.0	0.8	0.8853	0.9164
26	5.2	0.8	0.8981	0.9046
27	5.4	0.8	0.8749	0.8603
28	5.6	0.8	0.8308	0.7920
29	5.8	0.8	0.7582	0.7097
30	6.0	0.8	0.6872	0.6246
31	6.2	0.9	0.6005	0.5480
32	6.4	0.9	0.5457	0.4906
33	6.6	0.9	0.4868	0.4619
34	6.8	0.9	0.4897	0.4681
35	7.0	0.9	0.4886	0.5106
36	7.2	0.9	0.5669	0.5844
37	7.4	0.9	0.6120	0.6776
38	7.6	0.9	0.7402	0.7732
39	7.8	0.9	0.7676	0.8530

Numerical solution to equation (4.1): $\frac{dy}{dx} + y^2 = Sin^2(x)$ with the boundary values $(x, y) = (0, 0)$ and $(8, 0.9033)$ using Newton's iterative method. Comparison with Euler solution with initial value $(x, y) = (0, 0)$ shows rapid convergence to Euler solution.

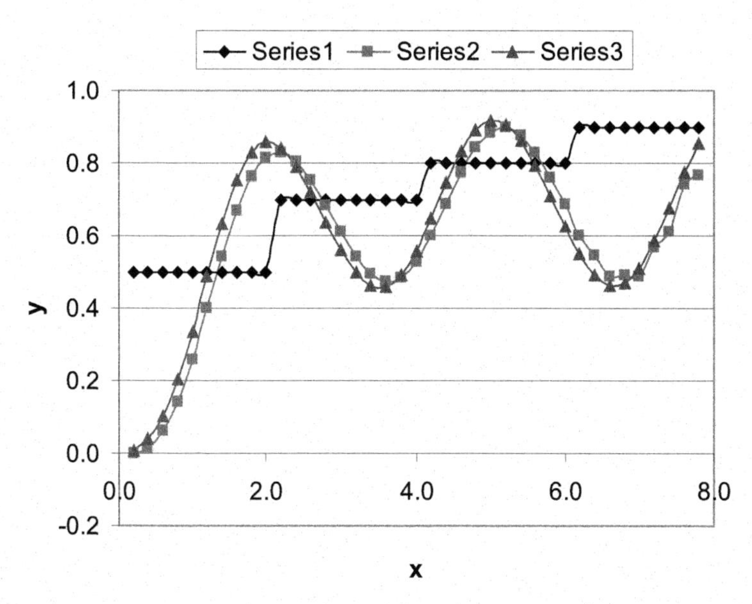

Figure 4.1 Numerical solution to equation (4.1): $\frac{dy}{dx} + y^2 = Sin^2(x)$ with the boundary values $(x, y) = (0, 0)$ and $(8, 0.9033)$ using Newton's iterative method. Showing the results of program number 4.2. Series 1 shows the initial values of y_i's taken, series 2 shows the results of Newton's method y_i's after 3rd iteration, and series 3 shows the Euler solution of Table 4.1. Rapid convergence to Euler solution is evident within 3rd iteration.

5

NUMERICAL SOLUTION OF BOUNDARY VALUE PROBLEM OF NON-LINEAR DIFFERENTIAL EQUATION

Example IV

In this chapter, we have presented the numerical solution to boundary value problem of the 4th non-linear differential equation in this book in comprehensive detail.

5.1 THE 4TH NON-LINEAR DIFFERENTIAL EQUATION IN THIS BOOK: EULER SOLUTION

In this chapter, we deal with the boundary value problem of the following non-linear differential equation:

$$\frac{dy}{dx} + y^3 = x \tag{5.1}$$

We first turn to Euler solution of this differential equation. To this end, we re-write equation (5.1) as

$$\frac{dy}{dx} = x - y^3 \tag{5.2}$$

which looks like equation (1.14): $\frac{dy}{dx} = f(x, y)$ which admits the iteration given by equation (1.20): $y_{n+1} = y_n + hf(x, y)$.

We have written program number 5.1 following Euler's method and solved equation (5.2) with initial values $(x, y) = (0, 0)$ with the numerical result shown in Table 5.1. From Table 5.1, we gather the boundary values $(x, y) = (0, 0)$ and $(8, 2.010)$.

DOI: 10.1201/9781003204916-5

TABLE 5.1
Euler Solution of Equation (5.2): $\frac{dy}{dx} = x - y^3$ with Initial Condition $(x, y) = (0, 0)$

i	x	y
1	0.2	0.0400
2	0.4	0.1200
3	0.6	0.2396
4	0.8	0.3969
5	1.0	0.5844
6	1.2	0.7845
7	1.4	0.9679
8	1.6	1.1066
9	1.8	1.1956
10	2.0	1.2538
11	2.2	1.2996
12	2.4	1.3406
13	2.6	1.3787
14	2.8	1.4146
15	3.0	1.4485
16	3.2	1.4807
17	3.4	1.5114
18	3.6	1.5409
19	3.8	1.5692
20	4.0	1.5964
21	4.2	1.6227
22	4.4	1.6481
23	4.6	1.6728
24	4.8	1.6966
25	5.0	1.7199
26	5.2	1.7424
27	5.4	1.7644
28	5.6	1.7858
29	5.8	1.8068
30	6.0	1.8272
31	6.2	1.8471
32	6.4	1.8667

(continued)

i	*x*	*y*
33	6.6	1.8858
34	6.8	1.9045
35	7.0	1.9229
36	7.2	1.9409
37	7.4	1.9586
38	7.6	1.9759
39	7.8	1.9930
40	8.0	2.0098

Program Number 5.1

```
h=0.2;
i=0;
x=0;
y=0;
Table[{i=i+1,x=x+h,y=y+h*(x-y^3)},{x,0,8,h}];
TableForm[%,TableSpacing->{2,2},TableHeadings->
{None,{"i","x","y"}}]

i=0;
x=0;
y=0;
p1=ListPlot[Table[{i=i+1;x=x+h,y=y+h*(x-y^3)},
{x,0,8,h}],
Frame->True,FrameLabel->{"x","y"},FrameTicks
->All,PlotStyle->{Black}]
```

5.2 THE 4TH NON-LINEAR DIFFERENTIAL EQUATION IN THIS BOOK: SOLUTION BY NEWTON'S ITERATIVE METHOD

For now we assume that the Euler solution is unknown and we have the non-linear differential equation (5.1): $\frac{dy}{dx} + y^3 = x$ with the boundary values $(x, y) = (0, 0)$ and $(8, 2.010)$.

We proceed by replacing the derivative dy/dx with finite difference approximation given by equation (1.12): $y' = (y_{i+1} - y_{i-1})/(2h)$. This results in the following difference equation:

$$\frac{y_{i+1} - y_{i-1}}{2h} + y_i^3 = x_i \text{ or, } -y_{i-1} + 2h(y_i^3 - x_i) + y_{i+1} = 0 \qquad (5.3)$$

We divide the interval $x = 0$ to 8 into 40 equal parts each being $h = 0.2$. We write equation (5.3) for $i = 1, 2, 3, \ldots, 39$ for which $x_i = 0.2, 0.4, 0.6, \ldots, 7.8$, respectively. We get the following system of non-linear algebraic equations:

$$
\begin{aligned}
f_1 &= -y_0 + 2h(y_1^3 - x_1) + y_2 = 0 \\
f_2 &= -y_1 + 2h(y_2^3 - x_2) + y_3 = 0 \\
f_3 &= -y_2 + 2h(y_3^3 - x_3) + y_4 = 0 \\
&\ldots \\
f_{39} &= -y_{38} + 2h(y_{39}^3 - x_{39}) + y_{40} = 0
\end{aligned}
$$

Here $y_0 = 0$ and $y_{40} = 2.010$ are constants. As such, the corresponding system of linear equations following Newton's iterative method is as follows:

$$
\begin{aligned}
6hy_1^2u_1 + 1u_2 &= -(-y_0 + 2h(y_1^3 - x_1) + y_2) \\
-1u_1 + 6hy_2^2u_2 + 1u_3 &= -(-y_1 + 2h(y_2^3 - x_2) + y_3) \\
-1u_2 + 6hy_3^2u_3 + 1u_4 &= -(-y_2 + 2h(y_3^3 - x_3) + y_4) \\
&\ldots \\
-1u_{38} + 6hy_{39}^2u_{39} &= -(-y_{38} + 2h(y_{39}^3 - x_{39}) + y_{40})
\end{aligned}
$$

To solve these equations which are linear in u_i's, we have taken initial set of values of y[i]'s as

```
y[0]=0;y[1]=1.;y[2]=1.;y[3]=1.;y[4]=1.;y[5]=1.;
y[6]=1.;y[7]=1.;y[8]=1.;y[9]=1.;
y[10]=1.;y[11]=1.;y[12]=1.;y[13]=1.;y[14]=1.;
y[15]=1.;y[16]=1.;y[17]=1.;
y[18]=1.;y[19]=1.;y[20]=1.;y[21]=2;y[22]=2;
y[23]=2;y[24]=2;y[25]=2;
y[26]=2;y[27]=2;y[28]=2;y[29]=2;y[30]=2;y[31]=2;
y[32]=2;y[33]=2;
y[34]=2;y[35]=2;y[36]=2;y[37]=2;y[38]=2;y[39]=2;
y[40]=2.010;
```

in program number 5.2. In using program number 5.2, the content of the program between the lines marked as

```
aaaaaaaaaaaaaaaaaaaaaaaaaaaaaaaaaaaaaaaa
```

and

bbbbbbbbbbbbbbbbbbbbbbbbbbbbbbbbbbbbbb

should be pasted $p-1$ times, after the line

bbbbbbbbbbbbbbbbbbbbbbbbbbbbbbbbbbbbbb

where p (= 6 in this case) is the number of iterations required to produce convergence of the values of y[i]'s. **The results of the program are shown in Table 5.2 and Figure 5.1.**

Program Number 5.2

```
h=0.2;
y[0]=0;
y[40]=2.010;

y[1]=1.;y[2]=1.;y[3]=1.;y[4]=1.;y[5]=1.;y[6]=1.;
y[7]=1.;y[8]=1.;y[9]=1.;y[10]=1.;
y[11]=1.;y[12]=1.;y[13]=1.;y[14]=1.;y[15]=1.;
y[16]=1.;y[17]=1.;y[18]=1.;y[19]=1.;y[20]=1.;y[21]=2;
y[22]=2;y[23]=2;y[24]=2;y[25]=2;y[26]=2;y[27]=2;
y[28]=2;y[29]=2;y[30]=2;
y[31]=2;y[32]=2;y[33]=2;y[34]=2;y[35]=2;y[36]=2;
y[37]=2;y[38]=2;y[39]=2;

x=0;
i=0;
Table[{i=i+1,x=x+h,y[i]},{i,0,38,1}];
TableForm[%,TableSpacing->{3,3},TableHeadings->{
None,{"i","x","y"}}]
x=0;
i=0;
p1=ListPlot[Table[{{i=i+1;x=x+h,
y[i]}},{i,0,38,1}],Frame->True,
FrameLabel->{"x","y"},FrameTicks->All,PlotStyle->
{Black},PlotRange->Automatic]

aaaaaaaaaaaaaaaaaaaaaaaaaaaaaaaaaaaa
G=NSolve[{
6*h*y[1]^2*u1+1*u2==-(-y[0]+2*h*(y[1]^3-0.1)+y[2]),
-1*u1+6*h*y[2]^2*u2+1*u3==-(-y[1]+2*h*(y[2]^3-0.2)+
y[3]),
-1*u2+6*h*y[3]^2*u3+1*u4==-(-y[2]+2*h*(y[3]^3-0.3)+
y[4]),
```

```
-1*u3+6*h*y[4]^2*u4+1*u5==-(-y[3]+2*h*(y[4]^3-
0.4)+y[5]),
-1*u4+6*h*y[5]^2*u5+1*u6==-(-y[4]+2*h*(y[5]^3-
0.5)+y[6]),
-1*u5+6*h*y[6]^2*u6+1*u7==-(-y[5]+2*h*(y[6]^3-
0.6)+y[7]),
-1*u6+6*h*y[7]^2*u7+1*u8==-(-y[6]+2*h*(y[7]^3-
0.7)+y[8]),
-1*u7+6*h*y[8]^2*u8+1*u9==-(-y[7]+2*h*(y[8]^3-
0.8)+y[9]),
-1*u8+6*h*y[9]^2*u9+1*u10==-(-y[8]+2*h*(y[9]^3-
0.9)+y[10]),
-1*u9+6*h*y[10]^2*u10+1*u11==-(-y[9]+2*h*(y[10]^3-
1.0)+y[11]),
-1*u10+6*h*y[11]^2*u11+1*u12==-(-y[10]+2*h*(y[11]^3-
1.1)+y[12]),
-1*u11+6*h*y[12]^2*u12+1*u13==-(-y[11]+2*h*(y[12]^3-
1.2)+y[13]),
-1*u12+6*h*y[13]^2*u13+1*u14==-(-y[12]+2*h*(y[13]^3-
1.3)+y[14]),
-1*u13+6*h*y[14]^2*u14+1*u15==-(-y[13]+2*h*(y[14]^3-
1.4)+y[15]),
-1*u14+6*h*y[15]^2*u15+1*u16==-(-y[14]+2*h*(y[15]^3-
1.5)+y[16]),
-1*u15+6*h*y[16]^2*u16+1*u17==-(-y[15]+2*h*(y[16]^3-
1.6)+y[17]),
-1*u16+6*h*y[17]^2*u17+1*u18==-(-y[16]+2*h*(y[17]^3-
1.7)+y[18]),
-1*u17+6*h*y[18]^2*u18+1*u19==-(-y[17]+2*h*(y[18]^3-
1.8)+y[19]),
-1*u18+6*h*y[19]^2*u19+1*u20==-(-y[18]+2*h*(y[19]^3-
1.9)+y[20]),
-1*u19+6*h*y[20]^2*u20+1*u21==-(-y[19]+2*h*(y[20]^3-
2.0)+y[21]),
-1*u20+6*h*y[21]^2*u21+1*u22==-(-y[20]+2*h*(y[21]^3-
2.1)+y[22]),
-1*u21+6*h*y[22]^2*u22+1*u23==-(-y[21]+2*h*(y[22]^3-
2.2)+y[23]),
-1*u22+6*h*y[23]^2*u23+1*u24==-(-y[22]+2*h*(y[23]^3-
2.3)+y[24]),
-1*u23+6*h*y[24]^2*u24+1*u25==-(-y[23]+2*h*(y[24]^3-
2.4)+y[25]),
-1*u24+6*h*y[25]^2*u25+1*u26==-(-y[24]+2*h*(y[25]^3-
2.5)+y[26]),
```

```
-1*u25+6*h*y[26]^2*u26+1*u27==-(-y[25]+2*h*(y[26]^3-
2.6)+y[27]),
-1*u26+6*h*y[27]^2*u27+1*u28==-(-y[26]+2*h*(y[27]^3-
2.7)+y[28]),
-1*u27+6*h*y[28]^2*u28+1*u29==-(-y[27]+2*h*(y[28]^3-
2.8)+y[29]),
-1*u28+6*h*y[29]^2*u29+1*u30==-(-y[28]+2*h*(y[29]^3-
2.9)+y[30]),
-1*u29+6*h*y[30]^2*u30+1*u31==-(-y[29]+2*h*(y[30]^3-
3.0)+y[31]),
-1*u30+6*h*y[31]^2*u31+1*u32==-(-y[30]+2*h*(y[31]^3-
3.1)+y[32]),
-1*u31+6*h*y[32]^2*u32+1*u33==-(-y[31]+2*h*(y[32]^3-
3.2)+y[33]),
-1*u32+6*h*y[33]^2*u33+1*u34==-(-y[32]+2*h*(y[33]^3-
3.3)+y[34]),
-1*u33+6*h*y[34]^2*u34+1*u35==-(-y[33]+2*h*(y[34]^3-
3.4)+y[35]),
-1*u34+6*h*y[35]^2*u35+1*u36==-(-y[34]+2*h*(y[35]^3-
3.5)+y[36]),
-1*u35+6*h*y[36]^2*u36+1*u37==-(-y[35]+2*h*(y[36]^3-
3.6)+y[37]),
-1*u36+6*h*y[37]^2*u37+1*u38==-(-y[36]+2*h*(y[37]^3-
3.7)+y[38]),
-1*u37+6*h*y[38]^2*u38+1*u39==-(-y[37]+2*h*(y[38]^3-
3.8)+y[39]),
-1*u38+6*h*y[39]^2*u39==-(-y[38]+2*h*(y[39]^3-
3.9)+y[40])},
{u1,u2,u3,u4,u5,u6,u7,u8,u9,u10,u11,u12,u13,u14,u15,
u16,u17,u18,u19,u20,
u21,u22,u23,u24,u25,u26,u27,u28,u29,u30,u31,u32,u33,
u34,u35,u36,u37,u38,u39}];

f[1]=N[EL1=Part[G,1];L1=u1/.EL1];
f[2]=N[EL2=Part[G,1];L2=u2/.EL2];
f[3]=N[EL3=Part[G,1];L3=u3/.EL3];
f[4]=N[EL4=Part[G,1];L4=u4/.EL4];
f[5]=N[EL5=Part[G,1];L5=u5/.EL5];
f[6]=N[EL6=Part[G,1];L6=u6/.EL6];
f[7]=N[EL7=Part[G,1];L7=u7/.EL7];
f[8]=N[EL8=Part[G,1];L8=u8/.EL8];
f[9]=N[EL9=Part[G,1];L9=u9/.EL9];
f[10]=N[EL10=Part[G,1];L10=u10/.EL10];
f[11]=N[EL11=Part[G,1];L11=u11/.EL11];
```

```
f[12]=N[EL12=Part[G,1];L12=u12/.EL12];
f[13]=N[EL13=Part[G,1];L13=u13/.EL13];
f[14]=N[EL14=Part[G,1];L14=u14/.EL14];
f[15]=N[EL15=Part[G,1];L15=u15/.EL15];
f[16]=N[EL16=Part[G,1];L16=u16/.EL16];
f[17]=N[EL17=Part[G,1];L17=u17/.EL17];
f[18]=N[EL18=Part[G,1];L18=u18/.EL18];
f[19]=N[EL19=Part[G,1];L19=u19/.EL19];
f[20]=N[EL20=Part[G,1];L20=u20/.EL20];
f[21]=N[EL21=Part[G,1];L21=u21/.EL21];
f[22]=N[EL22=Part[G,1];L22=u22/.EL22];
f[23]=N[EL23=Part[G,1];L23=u23/.EL23];
f[24]=N[EL24=Part[G,1];L24=u24/.EL24];
f[25]=N[EL25=Part[G,1];L25=u25/.EL25];
f[26]=N[EL26=Part[G,1];L26=u26/.EL26];
f[27]=N[EL27=Part[G,1];L27=u27/.EL27];
f[28]=N[EL28=Part[G,1];L28=u28/.EL28];
f[29]=N[EL29=Part[G,1];L29=u29/.EL29];
f[30]=N[EL30=Part[G,1];L30=u30/.EL30];
f[31]=N[EL31=Part[G,1];L31=u31/.EL31];
f[32]=N[EL32=Part[G,1];L32=u32/.EL32];
f[33]=N[EL33=Part[G,1];L33=u33/.EL33];
f[34]=N[EL34=Part[G,1];L34=u34/.EL34];
f[35]=N[EL35=Part[G,1];L35=u35/.EL35];
f[36]=N[EL36=Part[G,1];L36=u36/.EL36];
f[37]=N[EL37=Part[G,1];L37=u37/.EL37];
f[38]=N[EL38=Part[G,1];L38=u38/.EL38];
f[39]=N[EL39=Part[G,1];L39=u39/.EL39];

y[1]=f[1]+y[1];
y[2]=f[2]+y[2];
y[3]=f[3]+y[3];
y[4]=f[4]+y[4];y[5]=f[5]+y[5];
y[6]=f[6]+y[6];y[7]=f[7]+y[7];y[8]=f[8]+y[8];
y[9]=f[9]+y[9];y[10]=f[10]+y[10];
y[11]=f[11]+y[11];y[12]=f[12]+y[12];
y[13]=f[13]+y[13];y[14]=f[14]+y[14];
y[15]=f[15]+y[15];y[16]=f[16]+y[16];
y[17]=f[17]+y[17];y[18]=f[18]+y[18];
y[19]=f[19]+y[19];y[20]=f[20]+y[20];
y[21]=f[21]+y[21];y[22]=f[22]+y[22];
y[23]=f[23]+y[23];y[24]=f[24]+y[24];
y[25]=f[25]+y[25];y[26]=f[26]+y[26];
y[27]=f[27]+y[27];y[28]=f[28]+y[28];
```

```
y[29]=f[29]+y[29];y[30]=f[30]+y[30];
y[31]=f[31]+y[31];y[32]=f[32]+y[32];
y[33]=f[33]+y[33];y[34]=f[34]+y[34];
y[35]=f[35]+y[35];y[36]=f[36]+y[36];
y[37]=f[37]+y[37];y[38]=f[38]+y[38];
y[39]=f[39]+y[39];

x=0;
i=0;
Table[{i=i+1,x=x+h,y[i]},{i,0,38,1}];
TableForm[%,TableSpacing->{3,3},TableHeadings
->{None,{"i","x","y"}}]
x=0;
i=0;
p1=ListPlot[Table[{{i=i+1;x=x+h,
y[i]}},{i,0,38,1}],Frame->True,
FrameLabel->{"x","y"},FrameTicks->All,PlotStyle->
{Black},PlotRange->Automatic]
bbbbbbbbbbbbbbbbbbbbbbbbbbbbbbbbbbbb
```

TABLE 5.2
The Results of the Program Number 5.2

i	x	Initial values of y_i's taken	y_i's after 6th iteration	Euler solution of Table 5.1
1	0.2	1	0.0100	0.0400
2	0.4	1	0.0400	0.1200
3	0.6	1	0.0900	0.2396
4	0.8	1	0.1597	0.3969
5	1.0	1	0.2483	0.5844
6	1.2	1	0.3536	0.7845
7	1.4	1	0.4707	0.9679
8	1.6	1	0.5919	1.1066
9	1.8	1	0.7077	1.1956
10	2.0	1	0.8101	1.2538
11	2.2	1	0.8951	1.2996
12	2.4	1	0.9632	1.3406
13	2.6	1	1.0176	1.3787

(continued)

TABLE 5.2 (Continued)

i	x	Initial values of y_i's taken	y_i's after 6th iteration	Euler solution of Table 5.1
14	2.8	1	1.0618	1.4146
15	3.0	1	1.0988	1.4485
16	3.2	1	1.1311	1.4807
17	3.4	1	1.1600	1.5114
18	3.6	1	1.1867	1.5409
19	3.8	1	1.2116	1.5692
20	4.0	1	1.2352	1.5964
21	4.2	2	1.2578	1.6227
22	4.4	2	1.2794	1.6481
23	4.6	2	1.3002	1.6728
24	4.8	2	1.3202	1.6966
25	5.0	2	1.3397	1.7199
26	5.2	2	1.3585	1.7424
27	5.4	2	1.3768	1.7644
28	5.6	2	1.3946	1.7858
29	5.8	2	1.4119	1.8068
30	6.0	2	1.4288	1.8272
31	6.2	2	1.4452	1.8471
32	6.4	2	1.4614	1.8667
33	6.6	2	1.4769	1.8858
34	6.8	2	1.4929	1.9045
35	7.0	2	1.5060	1.9229
36	7.2	2	1.5267	1.9409
37	7.4	2	1.5226	1.9586
38	7.6	2	1.5949	1.9759
39	7.8	2	1.4198	1.9930

Numerical solution to equation (5.1): $\frac{dy}{dx} + y^3 = x$ with the boundary values $(x, y) = (0, 0)$ and $(8, 2.010)$ using Newton's iterative method. Comparison with Euler solution with initial value $(x, y) = (0, 0)$ shows rapid convergence to Euler solution.

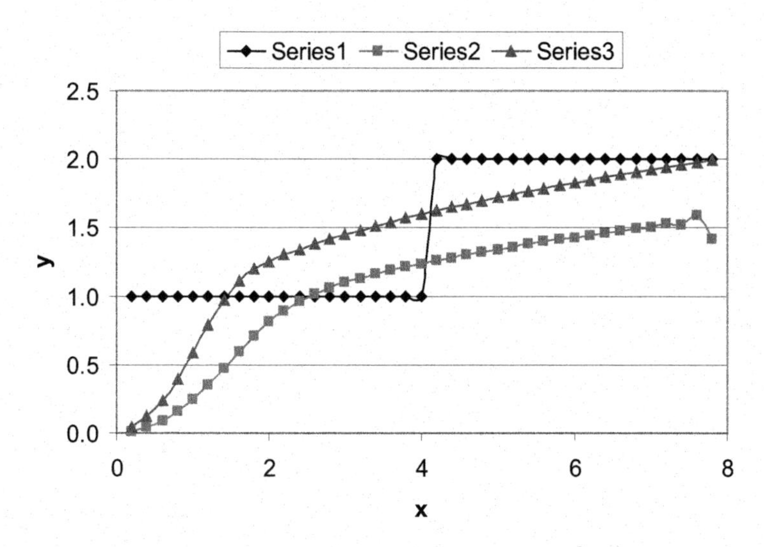

Figure 5.1 Numerical solution to equation (5.1): $\frac{dy}{dx} + y^3 = x$ with the boundary values $(x, y) = (0, 0)$ and $(8, 2.010)$ using Newton's iterative method. Showing the results of program number 5.2. Series 1 shows the initial values of y_i's taken, series 2 shows the results of Newton's method y_i's after 6th iteration, and series 3 shows the Euler solution of Table 5.1. Rapid convergence to Euler solution is evident. The two solutions agree within a multiplicative factor.

6

NUMERICAL SOLUTION OF BOUNDARY VALUE PROBLEM OF NON-LINEAR DIFFERENTIAL EQUATION

Example V

In this chapter, we have presented the numerical solution to boundary value problem of the 5th non-linear differential equation in this book in comprehensive detail.

6.1 THE 5TH NON-LINEAR DIFFERENTIAL EQUATION IN THIS BOOK: EULER SOLUTION

In this chapter, we deal with the boundary value problem of the following non-linear differential equation:

$$\frac{d^2y}{dx^2} + y^2 = x \tag{6.1}$$

We first turn to Euler solution of this differential equation. To this end, we re-write equation (6.1) as

$$\frac{d^2y}{dx^2} = x - y^2 \tag{6.2}$$

or,

$$\frac{dV}{dx} = x - y^2 \tag{6.3}$$

where

$$\frac{dy}{dx} = V \tag{6.4}$$

DOI: 10.1201/9781003204916-6

Both equations (6.3) and (6.4) look like equation (1.14): $\frac{dy}{dx} = f(x, y)$ which admits the iteration given by equation (1.20): $y_{n+1} = y_n + h\, f(x, y)$. As such, we obtained solution to equation (6.2) in two steps. In the first step, we solve equation (6.3) to obtain Euler solution for V and in the second step, we solve equation (6.4) to obtain Euler solution for y.

We have written program number 6.1 following Euler's method and solved equation (6.2) using initial values $(x, V) = (0, 3)$ and $(x, y) = (0, 0)$ with the numerical result shown in Table 6.1. From Table 6.1, we gather the boundary values $(x, y) = (0, 0)$ and $(8, 3.8546)$.

TABLE 6.1
Euler Solution of Equation (6.2): $\frac{d^2y}{dx^2} = x - y^2$ with Initial Condition $(x, V) = (0, 3)$ and $(x, y) = (0, 0)$ where $\frac{dy}{dx} = V$

i	*x*	*V*	*y*
1	0.2	3.0400	0.6080
2	0.4	3.0461	1.2172
3	0.6	2.8698	1.7912
4	0.8	2.3881	2.2688
5	1.0	1.5586	2.5805
6	1.2	0.4668	2.6739
7	1.4	−0.6831	2.5373
8	1.6	−1.6506	2.2071
9	1.8	−2.2649	1.7541
10	2.0	−2.4803	1.2581
11	2.2	−2.3569	0.7867
12	2.4	−2.0006	0.3866
13	2.6	−1.5105	0.0845
14	2.8	−0.9520	−0.1059
15	3.0	−0.3542	−0.1768
16	3.2	0.2795	−0.1208

(continued)

i	*x*	*V*	*y*
17	3.4	0.9566	0.0705
18	3.6	1.6756	0.4056
19	3.8	2.4027	0.8861
20	4.0	3.0457	1.4953
21	4.2	3.4385	2.1830
22	4.4	3.3654	2.8561
23	4.6	2.6540	3.3869
24	4.8	1.3198	3.6508
25	5.0	−0.3459	3.5817
26	5.2	−1.8715	3.2073
27	5.4	−2.8490	2.6376
28	5.6	−3.1203	2.0135
29	5.8	−2.7711	1.4593
30	6.0	−1.9970	1.0599
31	6.2	−0.9817	0.8635
32	6.4	0.1492	0.8934
33	6.6	1.3096	1.1553
34	6.8	2.4026	1.6358
35	7.0	3.2675	2.2893
36	7.2	3.6593	3.0212
37	7.4	3.3138	3.6839
38	7.6	2.1196	4.1078
39	7.8	0.3047	4.1688
40	8.0	−1.5710	3.8546

Program Number 6.1

```
i=0;
h=0.2;
x=0;
V=3;
y=0;
Table[{i=i+1,x=x+h,V=V+h*(x-y^2),y=y+h*(V)},
{x,0,7.8,h}];
TableForm[%,TableSpacing->{2,2},
TableHeadings->{None,{"i","x","V","y"}}]
```

```
i=0;
x=0;
V=3;
y=0;
p1=ListPlot[Table[{{x=x+h,V=V+h*(x-y^2)},
{x=x+h,y=y+h*(V)}},{x,0,7.8,h}],Frame->True,
FrameLabel->{"x","y or V"},FrameTicks->All,
PlotStyle->{Black}]

i=0;
x=0;
V=3;
y=0;
p2=ListPlot[Table[{{x=x+h,V=V+h*(x-y^2)};
{x=x+h,y=y+h*(V)}},{x,0,7.8,h}],Frame->True,
FrameLabel->{"x","y or V"},FrameTicks->All,
PlotStyle->{Black}]
```

6.2 THE 5TH NON-LINEAR DIFFERENTIAL EQUATION IN THIS BOOK: SOLUTION BY NEWTON'S ITERATIVE METHOD

For now we assume that the Euler solution is unknown and we have the non-linear differential equation (6.1): $\frac{d^2y}{dx^2}+y^2=x$ with the boundary values $(x, y) = (0, 0)$ and $(8, 3.8546)$.

We proceed by replacing the derivative d^2y/dx^2 with finite difference approximation given by equation (1.13): $y'' = (y_{i-1} - 2\,y_i + y_{i+1})/h^2$. This results in the following difference equation:

$$\frac{y_{i-1}-2y_i+y_{i+1}}{h^2}+y_i^2=x_i \;\; or, \;\; y_{i-1}-2y_i+h^2y_i^2+y_{i+1}-h^2x_i=0 \quad (6.5)$$

We divide the interval $x = 0$ to 8 into 40 equal parts each being $h = 0.2$. We write equation (6.5) for $i = 1, 2, 3, \ldots, 39$ for which $x_i = 0.2, 0.4, 0.6, \ldots, 7.8$, respectively. We get the following system of non-linear algebraic equations:

$$f_1 = y_0 - 2y_1 + h^2y_1^2 + y_2 - h^2x_1 = 0$$
$$f_2 = y_1 - 2y_2 + h^2y_2^2 + y_3 - h^2x_2 = 0$$

$$f_3 = y_2 - 2y_3 + h^2 y_3^2 + y_4 - h^2 x_3 = 0$$

$$\ldots$$

$$f_{39} = y_{38} - 2y_{39} + h^2 y_{39}^2 + y_{40} - h^2 x_{39} = 0$$

Here $y_0 = 0$ and $y_{40} = 3.8546$ are constants. As such, the corresponding system of linear equations following Newton's iterative method is as follows:

$$-2u_1 + 2h^2 y_1 u_1 + 1u_2 = -(y_0 - 2y_1 + h^2 y_1^2 + y_2 - h^2 x_1)$$
$$1u_1 - 2u_2 + 2h^2 y_2 u_2 + 1u_3 = -(y_1 - 2y_2 + h^2 y_2^2 + y_3 - h^2 x_2)$$
$$1u_2 - 2u_3 + 2h^2 y_3 u_3 + 1u_4 = -(y_2 - 2y_3 + h^2 y_3^2 + y_4 - h^2 x_3)$$
$$\ldots$$
$$1u_{38} - 2u_{39} + 2h^2 y_{39} u_{39} = -(y_{38} - 2y_{39} + h^2 y_{39}^2 + y_{40} - h^2 x_{39})$$

To solve these equations which are linear in u_i's, we have taken initial set of values of y[i]'s as

```
y[0]=0; y[1]=3;y[2]=2.5;y[3]=2;y[4]=2;y[5]=1.5;
y[6]=1;y[7]=2;y[8]=1;y[9]=1;y[10]=1;
y[11]=1;y[12]=1;y[13]=1;y[14]=1;y[15]=1;y[16]=1;
y[17]=1;y[18]=1;y[19]=1;y[20]=1;
y[21]=0.5;y[22]=0.5;y[23]=2.5;y[24]=0.5;y[25]=0.5;
y[26]=2.5;y[27]=0.5;y[28]=2.5;
y[29]=0.5;y[30]=0.5;y[31]=0.5;y[32]=0.5;y[33]=2.5;
y[34]=2.5;y[35]=3.5;y[36]=3.5;
y[37]=4;y[38]=3.5;y[39]=4;y[40]=3.8546;
```

in program number 6.2. In using program number 6.2, the content of the program between the lines marked as

```
aaaaaaaaaaaaaaaaaaaaaaaaaaaaaaaaaaaaaaa
```

and

```
bbbbbbbbbbbbbbbbbbbbbbbbbbbbbbbbbbbbbb
```

should be pasted p–1 times, after the line

```
bbbbbbbbbbbbbbbbbbbbbbbbbbbbbbbbbbbbbb
```

where p (= 6 in this case) is the number of iterations required to produce convergence of the values of y[i]'s. **The results of the program are shown in Table 6.2 and Figure 6.1.**

Program Number 6.2

```
h=0.2;
y[0]=0;
y[40]=3.8546;
y[1]=3;
y[2]=2.5;
y[3]=2;y[4]=2;y[5]=1.5;y[6]=1;y[7]=2;y[8]=1;
y[9]=1;y[10]=1;
y[11]=1;y[12]=1;y[13]=1;y[14]=1;y[15]=1;y[16]=1;
y[17]=1;y[18]=1;y[19]=1;y[20]=1;
y[21]=0.5;y[22]=0.5;y[23]=2.5;y[24]=0.5;y[25]=0.5;
y[26]=2.5;y[27]=0.5;y[28]=2.5;
y[29]=0.5;y[30]=0.5;y[31]=0.5;y[32]=0.5;y[33]=2.5;
y[34]=2.5;y[35]=3.5;y[36]=3.5;
y[37]=4;y[38]=3.5;y[39]=4;

x=0;
i=0;
Table[{i=i+1,x=x+h,y[i]},{i,0,38,1}];
TableForm[%,TableSpacing->{3,3},TableHeadings->{
None,{"i","x","y"}}]
x=0;
i=0;
p1=ListPlot[Table[{{i=i+1;x=x+h,
y[i]}},{i,0,38,1}],Frame->True,
FrameLabel->{"x","y"},FrameTicks->All,PlotStyle->
{Black},PlotRange->Automatic]

aaaaaaaaaaaaaaaaaaaaaaaaa
G=NSolve[{-2*u1+2*h^2*y[1]*u1+1*u2==-(y[0]-2*y[1]+h
^2*y[1]^2+y[2]-h^2*0.2),
1*u1-2*u2+2*h^2*y[2]*u2+1*u3==-(y[1]-2*y[2]+h^2*y[2
]^2+y[3]-h^2*0.4),
1*u2-2*u3+2*h^2*y[3]*u3+1*u4==-(y[2]-2*y[3]+h^2*y[3
]^2+y[4]-h^2*0.6),
1*u3-2*u4+2*h^2*y[4]*u4+1*u5==-(y[3]-2*y[4]+h^2*y[4
]^2+y[5]-h^2*0.8),
1*u4-2*u5+2*h^2*y[5]*u5+1*u6==-(y[4]-2*y[5]+h^2*y[5
]^2+y[6]-h^2*1.0),
1*u5-2*u6+2*h^2*y[6]*u6+1*u7==-(y[5]-2*y[6]+h^2*y[6
]^2+y[7]-h^2*1.2),
1*u6-2*u7+2*h^2*y[7]*u7+1*u8==-(y[6]-2*y[7]+h^2*y[7
```

```
]^2+y[8]-h^2*1.4),
1*u7-2*u8+2*h^2*y[8]*u8+1*u9==-(y[7]-2*y[8]+h^2*y[8
]^2+y[9]-h^2*1.6),
1*u8-2*u9+2*h^2*y[9]*u9+1*u10==-(y[8]-2*y[9]+h^2*y[
9]^2+y[10]-h^2*1.8),
1*u9-2*u10+2*h^2*y[10]*u10+1*u11==-(y[9]-2*y[10]+h^
2*y[10]^2+y[11]-h^2*2.0),
1*u10-2*u11+2*h^2*y[11]*u11+1*u12==-(y[10]-2*y[11]+h^
2*y[11]^2+y[12]-h^2*2.2),
1*u11-2*u12+2*h^2*y[12]*u12+1*u13==-(y[11]-2*y[12]+h^
2*y[12]^2+y[13]-h^2*2.4),
1*u12-2*u13+2*h^2*y[13]*u13+1*u14==-(y[12]-2*y[13]+h^
2*y[13]^2+y[14]-h^2*2.6),
1*u13-2*u14+2*h^2*y[14]*u14+1*u15==-(y[13]-2*y[14]+h^
2*y[14]^2+y[15]-h^2*2.8),
1*u14-2*u15+2*h^2*y[15]*u15+1*u16==-(y[14]-2*y[15]+h^
2*y[15]^2+y[16]-h^2*3.0),
1*u15-2*u16+2*h^2*y[16]*u16+1*u17==-(y[15]-2*y[16]+h^
2*y[16]^2+y[17]-h^2*3.2),
1*u16-2*u17+2*h^2*y[17]*u17+1*u18==-(y[16]-2*y[17]+h^
2*y[17]^2+y[18]-h^2*3.4),
1*u17-2*u18+2*h^2*y[18]*u18+1*u19==-(y[17]-2*y[18]+h^
2*y[18]^2+y[19]-h^2*3.6),
1*u18-2*u19+2*h^2*y[19]*u19+1*u20==-(y[18]-2*y[19]+h^
2*y[19]^2+y[20]-h^2*3.8),
1*u19-2*u20+2*h^2*y[20]*u20+1*u21==-(y[19]-2*y[20]+h^
2*y[20]^2+y[21]-h^2*4.0),
1*u20-2*u21+2*h^2*y[21]*u21+1*u22==-(y[20]-2*y[21]+h^
2*y[21]^2+y[22]-h^2*4.2),
1*u21-2*u22+2*h^2*y[22]*u22+1*u23==-(y[21]-2*y[22]+h^
2*y[22]^2+y[23]-h^2*4.4),
1*u22-2*u23+2*h^2*y[23]*u23+1*u24==-(y[22]-2*y[23]+h^
2*y[23]^2+y[24]-h^2*4.6),
1*u23-2*u24+2*h^2*y[24]*u24+1*u25==-(y[23]-2*y[24]+h^
2*y[24]^2+y[25]-h^2*4.8),
1*u24-2*u25+2*h^2*y[25]*u25+1*u26==-(y[24]-2*y[25]+h^
2*y[25]^2+y[26]-h^2*5.0),
1*u25-2*u26+2*h^2*y[26]*u26+1*u27==-(y[25]-2*y[26]+h^
2*y[26]^2+y[27]-h^2*5.2),
1*u26-2*u27+2*h^2*y[27]*u27+1*u28==-(y[26]-2*y[27]+h^
2*y[27]^2+y[28]-h^2*5.4),
1*u27-2*u28+2*h^2*y[28]*u28+1*u29==-(y[27]-2*y[28]+h^
2*y[28]^2+y[29]-h^2*5.6),
1*u28-2*u29+2*h^2*y[29]*u29+1*u30==-(y[28]-2*y[29]+h^
2*y[29]^2+y[30]-h^2*5.8),
1*u29-2*u30+2*h^2*y[30]*u30+1*u31==-(y[29]-2*y[30]+h^
```

```
2*y[30]^2+y[31]-h^2*6.0),
1*u30-2*u31+2*h^2*y[31]*u31+1*u32==-(y[30]-2*y[31]+h^
2*y[31]^2+y[32]-h^2*6.2),
1*u31-2*u32+2*h^2*y[32]*u32+1*u33==-(y[31]-2*y[32]+h^
2*y[32]^2+y[33]-h^2*6.4),
1*u32-2*u33+2*h^2*y[33]*u33+1*u34==-(y[32]-2*y[33]+h^
2*y[33]^2+y[34]-h^2*6.6),
1*u33-2*u34+2*h^2*y[34]*u34+1*u35==-(y[33]-2*y[34]+h^
2*y[34]^2+y[35]-h^2*6.8),
1*u34-2*u35+2*h^2*y[35]*u35+1*u36==-(y[34]-2*y[35]+h^
2*y[35]^2+y[36]-h^2*7.0),
1*u35-2*u36+2*h^2*y[36]*u36+1*u37==-(y[35]-2*y[36]+h^
2*y[36]^2+y[37]-h^2*7.2),
1*u36-2*u37+2*h^2*y[37]*u37+1*u38==-(y[36]-2*y[37]+h^
2*y[37]^2+y[38]-h^2*7.4),
1*u37-2*u38+2*h^2*y[38]*u38+1*u39==-(y[37]-
2*y[38]+h^2*y[38]^2+y[39]-h^2*7.6),
1*u38-2*u39+2*h^2*y[39]*u39 ==-(y[38]-2*y[39]+h^2*y
[39]^2+y[40]-h^2*7.8)},
{u1,u2,u3,u4,u5,u6,u7,u8,u9,u10,u11,u12,u13,u14,u15
,u16,u17,u18,u19,u20,
u21,u22,u23,u24,u25,u26,u27,u28,u29,u30,u31,u32,u33
,u34,u35,u36,u37,u38,u39}];

f[1]=N[EL1=Part[G,1];L1=u1/.EL1];
f[2]=N[EL2=Part[G,1];L2=u2/.EL2];
f[3]=N[EL3=Part[G,1];L3=u3/.EL3];
f[4]=N[EL4=Part[G,1];L4=u4/.EL4];
f[5]=N[EL5=Part[G,1];L5=u5/.EL5];
f[6]=N[EL6=Part[G,1];L6=u6/.EL6];
f[7]=N[EL7=Part[G,1];L7=u7/.EL7];
f[8]=N[EL8=Part[G,1];L8=u8/.EL8];
f[9]=N[EL9=Part[G,1];L9=u9/.EL9];
f[10]=N[EL10=Part[G,1];L10=u10/.EL10];
f[11]=N[EL11=Part[G,1];L11=u11/.EL11];
f[12]=N[EL12=Part[G,1];L12=u12/.EL12];
f[13]=N[EL13=Part[G,1];L13=u13/.EL13];
f[14]=N[EL14=Part[G,1];L14=u14/.EL14];
f[15]=N[EL15=Part[G,1];L15=u15/.EL15];
f[16]=N[EL16=Part[G,1];L16=u16/.EL16];
f[17]=N[EL17=Part[G,1];L17=u17/.EL17];
f[18]=N[EL18=Part[G,1];L18=u18/.EL18];
f[19]=N[EL19=Part[G,1];L19=u19/.EL19];
f[20]=N[EL20=Part[G,1];L20=u20/.EL20];
f[21]=N[EL21=Part[G,1];L21=u21/.EL21];
f[22]=N[EL22=Part[G,1];L22=u22/.EL22];
f[23]=N[EL23=Part[G,1];L23=u23/.EL23];
```

```
f[24]=N[EL24=Part[G,1];L24=u24/.EL24];
f[25]=N[EL25=Part[G,1];L25=u25/.EL25];
f[26]=N[EL26=Part[G,1];L26=u26/.EL26];
f[27]=N[EL27=Part[G,1];L27=u27/.EL27];
f[28]=N[EL28=Part[G,1];L28=u28/.EL28];
f[29]=N[EL29=Part[G,1];L29=u29/.EL29];
f[30]=N[EL30=Part[G,1];L30=u30/.EL30];
f[31]=N[EL31=Part[G,1];L31=u31/.EL31];
f[32]=N[EL32=Part[G,1];L32=u32/.EL32];
f[33]=N[EL33=Part[G,1];L33=u33/.EL33];
f[34]=N[EL34=Part[G,1];L34=u34/.EL34];
f[35]=N[EL35=Part[G,1];L35=u35/.EL35];
f[36]=N[EL36=Part[G,1];L36=u36/.EL36];
f[37]=N[EL37=Part[G,1];L37=u37/.EL37];
f[38]=N[EL38=Part[G,1];L38=u38/.EL38];
f[39]=N[EL39=Part[G,1];L39=u39/.EL39];

y[1]=f[1]+y[1];
y[2]=f[2]+y[2];
y[3]=f[3]+y[3];
y[4]=f[4]+y[4];y[5]=f[5]+y[5];
y[6]=f[6]+y[6];y[7]=f[7]+y[7];y[8]=f[8]+y[8];y[9]=
f[9]+y[9];y[10]=f[10]+y[10];
y[11]=f[11]+y[11];y[12]=f[12]+y[12];y[13]=f[13]+
y[13];y[14]=f[14]+y[14];
y[15]=f[15]+y[15];y[16]=f[16]+y[16];y[17]=f[17]+
y[17];y[18]=f[18]+y[18];
y[19]=f[19]+y[19];y[20]=f[20]+y[20];y[21]=f[21]+
y[21];y[22]=f[22]+y[22];
y[23]=f[23]+y[23];y[24]=f[24]+y[24];y[25]=f[25]+
y[25];y[26]=f[26]+y[26];
y[27]=f[27]+y[27];y[28]=f[28]+y[28];y[29]=f[29]+
y[29];y[30]=f[30]+y[30];
y[31]=f[31]+y[31];y[32]=f[32]+y[32];y[33]=f[33]+
y[33];y[34]=f[34]+y[34];
y[35]=f[35]+y[35];y[36]=f[36]+y[36];y[37]=f[37]+
y[37];y[38]=f[38]+y[38];
y[39]=f[39]+y[39];

x=0;
i=0;
Table[{i=i+1,x=x+h,y[i]},{i,0,38,1}];
TableForm[%,TableSpacing->{3,3},TableHead-
ings->{None,{"i","x","y"}}]
x=0;
i=0;
p1=ListPlot[Table[{{i=i+1;x=x+h,
```

```
y[i]}},{i,0,38,1}],Frame->True,
FrameLabel->{"x","y"},FrameTicks->All,PlotStyle->
{Black},PlotRange->Automatic]
bbbbbbbbbbbbbbbbbbbbbbbbbbbbbbbbbbbbbb
```

TABLE 6.2

The Results of the Program Number 6.2

i	x	Initial values of y_i's taken	y_i's after 6th iteration	Euler solution of Table 6.1
1	0.2	3.0	0.5806	0.6080
2	0.4	2.5	1.1557	1.2172
3	0.6	2.0	1.6933	1.7912
4	0.8	2.0	2.1403	2.2688
5	1.0	1.5	2.4361	2.5805
6	1.2	1.0	2.5344	2.6739
7	1.4	2.0	2.4239	2.5373
8	1.6	1.0	2.1344	2.2071
9	1.8	1.0	1.7266	1.7541
10	2.0	1.0	1.2716	1.2581
11	2.2	1.0	0.8319	0.7867
12	2.4	1.0	0.4525	0.3866
13	2.6	1.0	0.1610	0.0845
14	2.8	1.0	−0.0276	−0.1059
15	3.0	1.0	−0.1042	−0.1768
16	3.2	1.0	−0.0612	−0.1208
17	3.4	1.0	0.1097	0.0705
18	3.6	1.0	0.4161	0.4056
19	3.8	1.0	0.8596	0.8861
20	4.0	1.0	1.4255	1.4953
21	4.2	0.5	2.0701	2.1830
22	4.4	0.5	2.7114	2.8561
23	4.6	2.5	3.2345	3.3869
24	4.8	0.5	3.5232	3.6508
25	5.0	0.5	3.5074	3.5817
26	5.2	2.5	3.1996	3.2073
27	5.4	0.5	2.6902	2.6376

(continued)

i	x	Initial values of y_i's taken	y_i's after 6th iteration	Euler solution of Table 6.1
28	5.6	2.5	2.1074	2.0135
29	5.8	0.5	1.5709	1.4593
30	6.0	0.5	1.1677	1.0599
31	6.2	0.5	0.9500	0.8635
32	6.4	0.5	0.9442	0.8934
33	6.6	2.5	1.1588	1.1553
34	6.8	2.5	1.5836	1.6358
35	7.0	3.5	2.1802	2.2893
36	7.2	3.5	2.8666	3.0212
37	7.4	4.0	3.5123	3.6839
38	7.6	3.5	3.9606	4.1078
39	7.8	4.0	4.0854	4.1688

Numerical solution to equation (6.1): $\frac{d^2y}{dx^2} + y^2 = x$ with the boundary values $(x, y) = (0, 0)$ and $(8, 3.8546)$ using Newton's iterative method. Comparison with Euler solution with initial value $(x, y) = (0, 0)$ shows rapid convergence to Euler solution.

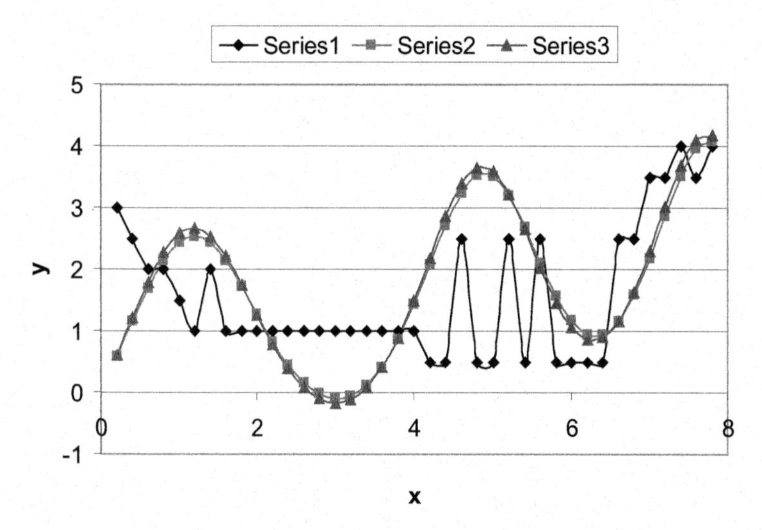

Figure 6.1 Numerical solution to equation (6.1): $\frac{d^2y}{dx^2} + y^2 = x$ with the boundary values $(x, y) = (0, 0)$ and $(8, 3.8546)$ using Newton's iterative method. Showing the results of program number 6.2. Series 1 shows the initial values of y_i's taken, series 2 shows the results of Newton's method y_i's after 6th iteration, and series 3 shows the Euler solution of Table 6.1. Rapid convergence to Euler solution is evident.

7

NUMERICAL SOLUTION OF BOUNDARY VALUE PROBLEM OF NON-LINEAR DIFFERENTIAL EQUATION

Example VI

In this chapter, we have presented the numerical solution to boundary value problem of the 6th non-linear differential equation in this book in comprehensive detail.

7.1 THE 6TH NON-LINEAR DIFFERENTIAL EQUATION IN THIS BOOK: EULER SOLUTION

In this chapter, we deal with the boundary value problem of the following non-linear differential equation:

$$\frac{d^2y}{dx^2} + y^2 = x^2 \tag{7.1}$$

We first turn to Euler solution of this differential equation. To this end, we re-write equation (7.1) as

$$\frac{d^2y}{dx^2} = x^2 - y^2 \tag{7.2}$$

or,

$$\frac{dV}{dx} = x^2 - y^2 \tag{7.3}$$

where

$$\frac{dy}{dx} = V \tag{7.4}$$

DOI: 10.1201/9781003204916-7

Both equations (7.3) and (7.4) look like equation (1.14): $\frac{dy}{dx} = f(x, y)$ which admits the iteration given by equation (1.20): $y_{n+1} = y_n + hf(x, y)$. As such, we obtained solution to equation (7.2) in two steps. In the first step, we solve equation (7.3) to obtain Euler solution for V and in the second step, we solve equation (7.4) to obtain Euler solution for y.

We have written program number 7.1 following Euler's method and solved equation (7.2) using initial values $(x, V) = (0, 3)$ and $(x, y) = (0, 0)$ with the numerical result shown in Table 7.1. From Table 7.1, we gather the boundary values $(x, y) = (0, 0)$ and $(8, 7.5376)$

Program Number 7.1

```
i=0;
h=0.2;
x=0;
V=3;
y=0;
Table[{i=i+1,x=x+h,V=V+h*(x^2-y^2),y=y+h*(V)},
{x,0,7.8,h}];
TableForm[%,TableSpacing->{2,2},TableHead-
ings->{None,{"i","x","V","y"}}]

i=0;
x=0;
V=3;
y=0;
p1=ListPlot[Table[{{x=x+h,V=V+h*(x^2-y^2)},
{x=x+h,y=y+h*(V)}},{x,0,7.8,h}],
Frame->True,FrameLabel->{"x","y or V"},Fra-
meTicks->All,PlotStyle->{Black}]
i=0;
x=0;
V=3;
y=0;
p2=ListPlot[Table[{{x=x+h,V=V+h*(x^2-y^2)};
{x=x+h,y=y+h*(V)}},{x,0,7.8,h}],
Frame->True,FrameLabel->{"x","y"},FrameTicks
->All,PlotStyle->{Black}]
```

TABLE 7.1
Euler Solution of Equation (7.2): $\frac{d^2y}{dx^2} = x^2 - y^2$ with Initial Condition $(x, V) = (0, 3)$ and $(x, y) = (0, 0)$ where $\frac{dy}{dx} = v$

i	*x*	*V*	*y*
1	0.2	3.0080	0.6016
2	0.4	2.9676	1.1951
3	0.6	2.7540	1.7459
4	0.8	2.2723	2.2004
5	1.0	1.5040	2.5012
6	1.2	0.5408	2.6093
7	1.4	−0.4289	2.5236
8	1.6	−1.1906	2.2854
9	1.8	−1.5872	1.9680
10	2.0	−1.5618	1.6556
11	2.2	−1.1420	1.4272
12	2.4	−0.3974	1.3477
13	2.6	0.5913	1.4660
14	2.8	1.7295	1.8119
15	3.0	2.8729	2.3865
16	3.2	3.7818	3.1428
17	3.4	4.1184	3.9665
18	3.6	3.5637	4.6793
19	3.8	2.0727	5.0938
20	4.0	0.0833	5.1105
21	4.2	−1.6120	4.7881
22	4.4	−2.3251	4.3230
23	4.6	−1.8308	3.9569
24	4.8	−0.3542	3.8860
25	5.0	1.6256	4.2112
26	5.2	3.4868	4.9085
27	5.4	4.5001	5.8085
28	5.6	4.0243	6.6134
29	5.8	2.0049	7.0144
30	6.0	−0.6354	6.8873

(continued)

Table 7.1 (Continued)

i	*x*	*V*	*y*
31	6.2	−2.4344	6.4004
32	6.4	−2.4354	5.9133
33	6.6	−0.7169	5.7700
34	6.8	1.8726	6.1445
35	7.0	4.1217	6.9688
36	7.2	4.7769	7.9242
37	7.4	3.1703	8.5582
38	7.6	0.0736	8.5730
39	7.8	−2.4575	8.0815
40	8.0	−2.7195	7.5376

7.2 THE 6TH NON-LINEAR DIFFERENTIAL EQUATION IN THIS BOOK: SOLUTION BY NEWTON'S ITERATIVE METHOD

For now we assume that the Euler solution is unknown and we have the non-linear differential equation (7.1): $\frac{d^2y}{dx^2} + y^2 = x^2$ with the boundary values $(x, y) = (0, 0)$ and $(8, 7.5376)$.

We proceed by replacing the derivative d^2y/dx^2 with finite difference approximation given by equation (1.13): $y'' = (y_{i-1} - 2\,y_i + y_{i+1})/h^2$. This results in the following difference equation:

$$\frac{y_{i-1} - 2y_i + y_{i+1}}{h^2} + y_i^2 = x_i^2 \text{ or, } y_{i-1} - 2y_i + h^2y_i^2 + y_{i+1} - h^2x_i^2 = 0 \quad (7.5)$$

We divide the interval $x = 0$ to 8 into 40 equal parts each being $h = 0.2$. We write equation (7.5) for $i = 1, 2, 3, \ldots, 39$ for which $x_i = 0.2, 0.4, 0.6, \ldots, 7.8$, respectively. We get the following system of non-linear algebraic equations:

$$f_1 = y_0 - 2y_1 + h^2y_1^2 + y_2 - h^2x_1^2 = 0$$
$$f_2 = y_1 - 2y_2 + h^2y_2^2 + y_3 - h^2x_2^2 = 0$$
$$f_3 = y_2 - 2y_3 + h^2y_3^2 + y_4 - h^2x_3^2 = 0$$
$$\ldots$$
$$f_{39} = y_{38} - 2y_{39} + h^2y_{39}^2 + y_{40} - h^2x_{39}^2 = 0$$

Here $y_0 = 0$ and $y_{40} = 7.5376$ are constants. As such, the corresponding system of linear equations following Newton's iterative method is as follows:

$$
\begin{aligned}
-2u_1 + 2h^2 y_1 u_1 + 1u_2 &= -(y_0 - 2y_1 + h^2 y_1^2 + y_2 - h^2 x_1^2) \\
1u_1 - 2u_2 + 2h^2 y_2 u_2 + 1u_3 &= -(y_1 - 2y_2 + h^2 y_2^2 + y_3 - h^2 x_2^2) \\
1u_2 - 2u_3 + 2h^2 y_3 u_3 + 1u_4 &= -(y_2 - 2y_3 + h^2 y_3^2 + y_4 - h^2 x_3^2) \\
&\;\;\ldots \\
1u_{38} - 2u_{39} + 2h^2 y_{39} u_{39} &= -(y_{38} - 2y_{39} + h^2 y_{39}^2 + y_{40} - h^2 x_{39}^2)
\end{aligned}
$$

To solve these equations which are linear in u_i's, we have taken initial set of values of y[i]'s as

```
y[0]=0;y[1]=1;y[2]=2;y[3]=2;y[4]=2;y[5]=1.5;
y[6]=1;y[7]=2;y[8]=1;y[9]=1;
y[10]=1;y[11]=3;y[12]=4;y[13]=3;y[14]=4;
y[15]=3;y[16]=4;y[17]=4;
y[18]=3;y[19]=3;y[20]=4;y[21]=5.5;y[22]=5.5;
y[23]=5.5;y[24]=5.5;
y[25]=5.5;y[26]=5.5;y[27]=5.5;y[28]=3.5;y[29]=7.5;
y[30]=7.5;y[31]=7.5;
y[32]=7.5;y[33]=7.5;y[34]=7.5;y[35]=7.5;y[36]=7.5;
y[37]=7;y[38]=7.5;
y[39]=7;y[40]=7.5376;
```

in program number 7.2. In using program number 7.2, the content of the program between the lines marked as

```
aaaaaaaaaaaaaaaaaaaaaaaaaaaaaaaaaaaaaa
```

and

```
bbbbbbbbbbbbbbbbbbbbbbbbbbbbbbbbbbbbbbb
```

should be pasted $p-1$ times, after the line

```
bbbbbbbbbbbbbbbbbbbbbbbbbbbbbbbbbbbbbbb
```

where p (= 5 in this case) is the number of iterations required to produce convergence of the values of y[i]'s. **The results of the program are shown in Table 7.2 and Figure 7.1.**

Program Number 7.2

```
h=0.2;
y[0]=0;
y[40]=7.5376;
y[1]=1;y[2]=2;y[3]=2;y[4]=2;y[5]=1.5;y[6]=1;
y[7]=2;y[8]=1;y[9]=1;y[10]=1;
y[11]=3;y[12]=4;y[13]=3;y[14]=4;y[15]=3;y[16]=4;
y[17]=4;y[18]=3;y[19]=3;y[20]=4;
y[21]=5.5;y[22]=5.5;y[23]=5.5;y[24]=5.5;y[25]=5.5;
y[26]=5.5;y[27]=5.5;y[28]=3.5;
y[29]=7.5;y[30]=7.5;y[31]=7.5;y[32]=7.5;y[33]=7.5;
y[34]=7.5;y[35]=7.5;y[36]=7.5;
y[37]=7;y[38]=7.5;y[39]=7;

x=0;
i=0;
Table[{i=i+1,x=x+h,y[i]},{i,0,38,1}];
TableForm[%,TableSpacing->{3,3},TableHead-
ings->{None,{"i","x","y"}}]

x=0;
i=0;
p1=ListPlot[Table[{{i=i+1;x=x+h,
y[i]}},{i,0,38,1}],Frame->True,
FrameLabel->{"x","y"},FrameTicks->All,PlotStyle->
{Black},PlotRange->Automatic]

aaaaaaaaaaaaaaaaaaaaaaaaaaaaaaaa
G=NSolve[{-2*u1+2*h^2*y[1]*u1+1*u2==-(y[0]-2*y[1]+h
^2*y[1]^2+y[2]-h^2*0.2^2),
1*u1-2*u2+2*h^2*y[2]*u2+1*u3==-(y[1]-2*y[2]+h^2*y[2]^
2+y[3]-h^2*0.4^2),
1*u2-2*u3+2*h^2*y[3]*u3+1*u4==-(y[2]-2*y[3]+h^2*y[3]^
2+y[4]-h^2*0.6^2),
1*u3-2*u4+2*h^2*y[4]*u4+1*u5==-(y[3]-2*y[4]+h^2*y[4]^
2+y[5]-h^2*0.8^2),
1*u4-2*u5+2*h^2*y[5]*u5+1*u6==-(y[4]-2*y[5]+h^2*y[5]^
2+y[6]-h^2*1.0^2),
1*u5-2*u6+2*h^2*y[6]*u6+1*u7==-(y[5]-2*y[6]+h^2*y[6]^
2+y[7]-h^2*1.2^2),
1*u6-2*u7+2*h^2*y[7]*u7+1*u8==-(y[6]-2*y[7]+h^2*y[7]^
2+y[8]-h^2*1.4^2),
1*u7-2*u8+2*h^2*y[8]*u8+1*u9==-(y[7]-2*y[8]+h^2*y[8]^
2+y[9]-h^2*1.6^2),
```

```
1*u8-2*u9+2*h^2*y[9]*u9+1*u10==-(y[8]-2*y[9]+h^2*y[9]
^2+y[10]-h^2*1.8^2),
1*u9-2*u10+2*h^2*y[10]*u10+1*u11==-(y[9]-2*y[10]+h^2*
y[10]^2+y[11]-h^2*2.0^2),
1*u10-2*u11+2*h^2*y[11]*u11+1*u12==-(y[10]-2*y[11]+h^
2*y[11]^2+y[12]-h^2*2.2^2),
1*u11-2*u12+2*h^2*y[12]*u12+1*u13==-(y[11]-2*y[12]+h^
2*y[12]^2+y[13]-h^2*2.4^2),
1*u12-2*u13+2*h^2*y[13]*u13+1*u14==-(y[12]-2*y[13]+h^
2*y[13]^2+y[14]-h^2*2.6^2),
1*u13-2*u14+2*h^2*y[14]*u14+1*u15==-(y[13]-2*y[14]+h^
2*y[14]^2+y[15]-h^2*2.8^2),
1*u14-2*u15+2*h^2*y[15]*u15+1*u16==-(y[14]-2*y[15]+h^
2*y[15]^2+y[16]-h^2*3.0^2),
1*u15-2*u16+2*h^2*y[16]*u16+1*u17==-(y[15]-2*y[16]+h^
2*y[16]^2+y[17]-h^2*3.2^2),
1*u16-2*u17+2*h^2*y[17]*u17+1*u18==-(y[16]-2*y[17]+h^
2*y[17]^2+y[18]-h^2*3.4^2),
1*u17-2*u18+2*h^2*y[18]*u18+1*u19==-(y[17]-2*y[18]+h^
2*y[18]^2+y[19]-h^2*3.6^2),
1*u18-2*u19+2*h^2*y[19]*u19+1*u20==-(y[18]-2*y[19]+h^
2*y[19]^2+y[20]-h^2*3.8^2),
1*u19-2*u20+2*h^2*y[20]*u20+1*u21==-(y[19]-2*y[20]+h^
2*y[20]^2+y[21]-h^2*4.0^2),
1*u20-2*u21+2*h^2*y[21]*u21+1*u22==-(y[20]-2*y[21]+h^
2*y[21]^2+y[22]-h^2*4.2^2),
1*u21-2*u22+2*h^2*y[22]*u22+1*u23==-(y[21]-2*y[22]+h^
2*y[22]^2+y[23]-h^2*4.4^2),
1*u22-2*u23+2*h^2*y[23]*u23+1*u24==-(y[22]-2*y[23]+h^
2*y[23]^2+y[24]-h^2*4.6^2),
1*u23-2*u24+2*h^2*y[24]*u24+1*u25==-(y[23]-2*y[24]+h^
2*y[24]^2+y[25]-h^2*4.8^2),
1*u24-2*u25+2*h^2*y[25]*u25+1*u26==-(y[24]-2*y[25]+h^
2*y[25]^2+y[26]-h^2*5.0^2),
1*u25-2*u26+2*h^2*y[26]*u26+1*u27==-(y[25]-2*y[26]+h^
2*y[26]^2+y[27]-h^2*5.2^2),
1*u26-2*u27+2*h^2*y[27]*u27+1*u28==-(y[26]-2*y[27]+h^
2*y[27]^2+y[28]-h^2*5.4^2),
1*u27-2*u28+2*h^2*y[28]*u28+1*u29==-(y[27]-2*y[28]+h^
2*y[28]^2+y[29]-h^2*5.6^2),
1*u28-2*u29+2*h^2*y[29]*u29+1*u30==-(y[28]-2*y[29]+h^
2*y[29]^2+y[30]-h^2*5.8^2),
1*u29-2*u30+2*h^2*y[30]*u30+1*u31==-(y[29]-2*y[30]+h^
2*y[30]^2+y[31]-h^2*6.0^2),
```

```
1*u30-2*u31+2*h^2*y[31]*u31+1*u32==-(y[30]-2*y[31]+h^
2*y[31]^2+y[32]-h^2*6.2^2),
1*u31-2*u32+2*h^2*y[32]*u32+1*u33==-(y[31]-2*y[32]+h^
2*y[32]^2+y[33]-h^2*6.4^2),
1*u32-2*u33+2*h^2*y[33]*u33+1*u34==-(y[32]-2*y[33]+h^
2*y[33]^2+y[34]-h^2*6.6^2),
1*u33-2*u34+2*h^2*y[34]*u34+1*u35==-(y[33]-2*y[34]+h^
2*y[34]^2+y[35]-h^2*6.8^2),
1*u34-2*u35+2*h^2*y[35]*u35+1*u36==-(y[34]-2*y[35]+h^
2*y[35]^2+y[36]-h^2*7.0^2),
1*u35-2*u36+2*h^2*y[36]*u36+1*u37==-(y[35]-2*y[36]+h^
2*y[36]^2+y[37]-h^2*7.2^2),
1*u36-2*u37+2*h^2*y[37]*u37+1*u38==-(y[36]-2*y[37]+h^
2*y[37]^2+y[38]-h^2*7.4^2),
1*u37-2*u38+2*h^2*y[38]*u38+1*u39==-(y[37]-2*y[38]+h^
2*y[38]^2+y[39]-h^2*7.6^2),
1*u38-2*u39+2*h^2*y[39]*u39==-(y[38]-2*y[39]+h^2*y
[39]^2+y[40]-h^2*7.8^2)},
{u1,u2,u3,u4,u5,u6,u7,u8,u9,u10,u11,u12,u13,u14,u15
,u16,u17,u18,u19,u20,
u21,u22,u23,u24,u25,u26,u27,u28,u29,u30,u31,u32,u33
,u34,u35,u36,u37,u38,u39}];

f[1]=N[EL1=Part[G,1];L1=u1/.EL1];
f[2]=N[EL2=Part[G,1];L2=u2/.EL2];
f[3]=N[EL3=Part[G,1];L3=u3/.EL3];
f[4]=N[EL4=Part[G,1];L4=u4/.EL4];
f[5]=N[EL5=Part[G,1];L5=u5/.EL5];
f[6]=N[EL6=Part[G,1];L6=u6/.EL6];
f[7]=N[EL7=Part[G,1];L7=u7/.EL7];
f[8]=N[EL8=Part[G,1];L8=u8/.EL8];
f[9]=N[EL9=Part[G,1];L9=u9/.EL9];
f[10]=N[EL10=Part[G,1];L10=u10/.EL10];
f[11]=N[EL11=Part[G,1];L11=u11/.EL11];
f[12]=N[EL12=Part[G,1];L12=u12/.EL12];
f[13]=N[EL13=Part[G,1];L13=u13/.EL13];
f[14]=N[EL14=Part[G,1];L14=u14/.EL14];
f[15]=N[EL15=Part[G,1];L15=u15/.EL15];
f[16]=N[EL16=Part[G,1];L16=u16/.EL16];
f[17]=N[EL17=Part[G,1];L17=u17/.EL17];
f[18]=N[EL18=Part[G,1];L18=u18/.EL18];
f[19]=N[EL19=Part[G,1];L19=u19/.EL19];
f[20]=N[EL20=Part[G,1];L20=u20/.EL20];
f[21]=N[EL21=Part[G,1];L21=u21/.EL21];
```

```
f[22]=N[EL22=Part[G,1];L22=u22/.EL22];
f[23]=N[EL23=Part[G,1];L23=u23/.EL23];
f[24]=N[EL24=Part[G,1];L24=u24/.EL24];
f[25]=N[EL25=Part[G,1];L25=u25/.EL25];
f[26]=N[EL26=Part[G,1];L26=u26/.EL26];
f[27]=N[EL27=Part[G,1];L27=u27/.EL27];
f[28]=N[EL28=Part[G,1];L28=u28/.EL28];
f[29]=N[EL29=Part[G,1];L29=u29/.EL29];
f[30]=N[EL30=Part[G,1];L30=u30/.EL30];
f[31]=N[EL31=Part[G,1];L31=u31/.EL31];
f[32]=N[EL32=Part[G,1];L32=u32/.EL32];
f[33]=N[EL33=Part[G,1];L33=u33/.EL33];
f[34]=N[EL34=Part[G,1];L34=u34/.EL34];
f[35]=N[EL35=Part[G,1];L35=u35/.EL35];
f[36]=N[EL36=Part[G,1];L36=u36/.EL36];
f[37]=N[EL37=Part[G,1];L37=u37/.EL37];
f[38]=N[EL38=Part[G,1];L38=u38/.EL38];
f[39]=N[EL39=Part[G,1];L39=u39/.EL39];

y[1]=f[1]+y[1];
y[2]=f[2]+y[2];
y[3]=f[3]+y[3];
y[4]=f[4]+y[4];y[5]=f[5]+y[5];
y[6]=f[6]+y[6];y[7]=f[7]+y[7];y[8]=f[8]+y[8];y[9]=
f[9]+y[9];y[10]=f[10]+y[10];
y[11]=f[11]+y[11];y[12]=f[12]+y[12];y[13]=f[13]+
y[13];y[14]=f[14]+y[14];
y[15]=f[15]+y[15];y[16]=f[16]+y[16];y[17]=f[17]+
y[17];y[18]=f[18]+y[18];
y[19]=f[19]+y[19];y[20]=f[20]+y[20];y[21]=f[21]+
y[21];y[22]=f[22]+y[22];
y[23]=f[23]+y[23];y[24]=f[24]+y[24];y[25]=f[25]+
y[25];y[26]=f[26]+y[26];
y[27]=f[27]+y[27];y[28]=f[28]+y[28];y[29]=f[29]+
y[29];y[30]=f[30]+y[30];
y[31]=f[31]+y[31];y[32]=f[32]+y[32];y[33]=f[33]+
y[33];y[34]=f[34]+y[34];
y[35]=f[35]+y[35];y[36]=f[36]+y[36];y[37]=f[37]+
y[37];y[38]=f[38]+y[38];
y[39]=f[39]+y[39];

x=0;
i=0;
Table[{i=i+1,x=x+h,y[i]},{i,0,38,1}];
```

```
TableForm[%,TableSpacing->{3,3},TableHead-
ings->{None,{"i","x","y"}}]
x=0;
i=0;
p1=ListPlot[Table[{{i=i+1;x=x+h,
y[i]}},{i,0,38,1}],Frame->True,
FrameLabel->{"x","y"},FrameTicks->All,PlotStyle->
{Black},PlotRange->Automatic]
```

bbbbbbbbbbbbbbbbbbbbb

TABLE 7.2
The Results of the Program Number 7.2

i	x	Initial values of y_i's taken	y_i's after 5th iteration	Euler solution of Table 7.1
1	0.2	1.0	0.5090	0.6016
2	0.4	2.0	1.0120	1.1951
3	0.6	2.0	1.4904	1.7459
4	0.8	2.0	1.9138	2.2004
5	1.0	1.5	2.2412	2.5012
6	1.2	1.0	2.4291	2.6093
7	1.4	2.0	2.4498	2.5236
8	1.6	1.0	2.3106	2.2854
9	1.8	1.0	2.0611	1.9680
10	2.0	1.0	1.7807	1.6556
11	2.2	3.0	1.5558	1.4272
12	2.4	4.0	1.4624	1.3477
13	2.6	3.0	1.5569	1.4660
14	2.8	4.0	1.8704	1.8119
15	3.0	3.0	2.3987	2.3865
16	3.2	4.0	3.0859	3.1428
17	3.4	4.0	3.8146	3.9665
18	3.6	3.0	4.4245	4.6793
19	3.8	3.0	4.7736	5.0938
20	4.0	4.0	4.8059	5.1105
21	4.2	5.5	4.5757	4.7881
22	4.4	5.5	4.2237	4.3230

(continued)

i	x	Initial values of y_i's taken	y_i's after 5th iteration	Euler solution of Table 7.1
23	4.6	5.5	3.9326	3.9569
24	4.8	5.5	3.8750	3.8860
25	5.0	5.5	4.1588	4.2112
26	5.2	5.5	4.7798	4.9085
27	5.4	5.5	5.5906	5.8085
28	5.6	3.5	6.3230	6.6134
29	5.8	7.5	6.7117	7.0144
30	6.0	7.5	6.6588	6.8873
31	6.2	7.5	6.2926	6.4004
32	6.4	7.5	5.8870	5.9133
33	6.6	7.5	5.7340	5.7700
34	6.8	7.5	6.0210	6.1445
35	7.0	7.5	6.7326	6.9688
36	7.2	7.5	7.6084	7.9242
37	7.4	7.0	8.2438	8.5582
38	7.6	7.5	8.3573	8.5730
39	7.8	7.0	8.0074	8.0815

Numerical solution to equation (7.1): $\frac{d^2y}{dx^2} + y^2 = x^2$ with the boundary values $(x, y) = (0, 0)$ and $(8, 7.5376)$ using Newton's iterative method. Comparison with Euler solution with initial value $(x, y) = (0, 0)$ shows rapid convergence to Euler solution.

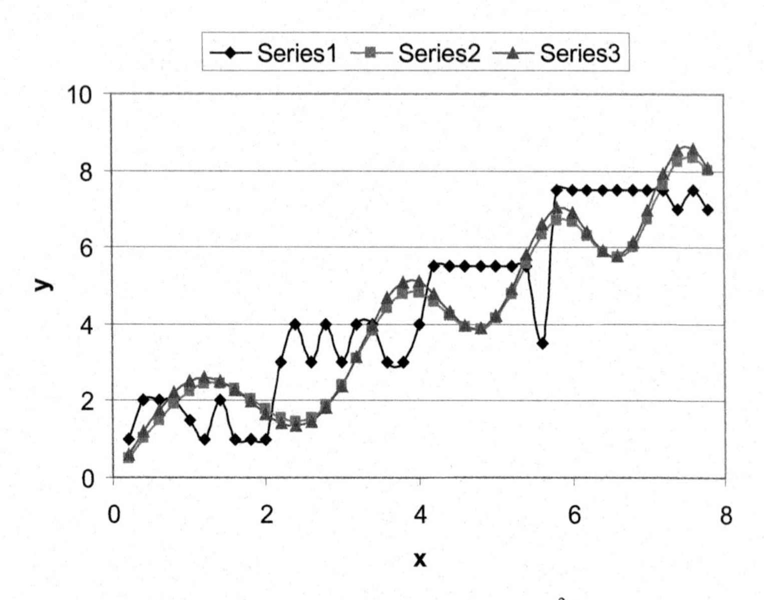

Figure 7.1 Numerical solution to equation (7.1): $\frac{d^2y}{dx^2} + y^2 = x^2$ with the boundary values (x, y) = (0, 0) and (8, 7.5376) using Newton's iterative method. Showing the results of program number 7.2. Series 1 shows the initial values of y_i's taken, series 2 shows the results of Newton's method y_i's after 5th iteration, and series 3 shows the Euler solution of Table 7.1. Rapid convergence to Euler solution is evident.

8

NUMERICAL SOLUTION OF BOUNDARY VALUE PROBLEM OF NON-LINEAR DIFFERENTIAL EQUATION

A Laborious Exercise

In this chapter, we have presented the numerical solution to boundary value problem of the 7th non-linear differential equation in this book in comprehensive detail.

8.1 THE 7TH NON-LINEAR DIFFERENTIAL EQUATION IN THIS BOOK: EULER SOLUTION

In this chapter, we deal with the boundary value problem of the following non-linear differential equation:

$$\frac{d^2y}{dx^2} = -\frac{y}{(y^2 + r^2(x))^{3/2}} \tag{8.1}$$

where

$$r = -c\ \mathrm{Cos}(x) + \frac{1}{2}c^2(1 - \mathrm{Cos}(2x)) \tag{8.2}$$

and c is a constant parameter, say 0.2.

We first turn to Euler solution of this differential equation. To this end, we re-write equation (8.1) as

$$\frac{dV}{dx} = -\frac{y}{(y^2 + r^2(x))^{3/2}} \tag{8.3}$$

where

$$\frac{dy}{dx} = V \tag{8.4}$$

DOI: 10.1201/9781003204916-8

Both equations (8.3) and (8.4) look like equation (1.14): $\frac{dy}{dx} = f(x, y)$ which admits the iteration given by equation (1.20): $y_{n+1} = y_n + hf(x, y)$. As such, we obtained solution to equation (8.1) in two steps. In the first step, we solve equation (8.3) to obtain Euler solution for V and in the second step, we solve equation (8.4) to obtain Euler solution for y.

We have written program number 8.1 following Euler's method and solved equation (8.1) using initial values $(x, V) = (0, 1)$ and $(x, y) = (0, 0)$ with the numerical result shown in Table 8.1. From Table 8.1, we gather the boundary values $(x, y) = (0, 0)$ and $(0.8, -0.0579)$.

Program Number 8.1

```
c=0.2;
h=0.02;
x=0;
y=0;

V=1;
i=0;
Table[{i=i+1,x=x+h,V=V+h*(-y/((y^2+(-c*
Cos[x]+0.5*c^2*(1-1*Cos[2*x]))^2)^1.5)),
y=y+h*(V)},{x,0,0.8,h}];
TableForm[%,TableSpacing->{2,2},TableHead-
ings->{None, {"i","x","V","y"}}]

x=0;
y=0;
V=1;
p2=ListPlot[Table[{{x=x+h,
V=V+h*(-y/((y^2+(-c*Cos[x]+0.5*c^2*(1-1*Cos[2
*x]))^2)^1.5))};
{x=x+h,y=y+h*(V)}},{x,0,0.8,h}],Frame->True,FrameL-
abel->{"x","y or V"},
FrameTicks->All,PlotStyle->{Black}]
```

8.2 THE 7TH NON-LINEAR DIFFERENTIAL EQUATION IN THIS BOOK: SOLUTION BY NEWTON'S ITERATIVE METHOD

For now we assume that the Euler solution is unknown and we have the non-linear differential equation (8.1): $\frac{d^2y}{dx^2} = -\frac{y}{(y^2 + r^2(x))^{3/2}}$ with the boundary values $(x, y) = (0, 0)$ and $(0.8, -0.0579)$.

We proceed by replacing the derivative d^2y/dx^2 with finite difference approximation given by equation (1.13): $y'' = (y_{i-1} - 2y_i + y_{i+1})/h^2$. This results in the following difference equation:

$$\frac{y_{i-1} - 2y_i + y_{i+1}}{h^2} + y_i(y_i^2 + r^2(x_i))^{-3/2} = 0$$

or,
$$y_{i-1} - 2y_i + h^2 y_i(y_i^2 + r^2(x_i))^{-3/2} + y_{i+1} = 0 \qquad (8.5)$$

We divide the interval $x = 0$ to 0.8 into 40 equal parts each being $h = 0.02$. We write equation (8.5) for $i = 1, 2, 3, \ldots, 39$ for which $x_i = 0.02, 0.04, 0.06, \ldots, 0.78$, respectively. We get the following system of non-linear algebraic equations:

$$f_1 = y_0 - 2y_1 + h^2 y_1(y_1^2 + r^2(x_1))^{-3/2} + y_2 = 0$$
$$f_2 = y_1 - 2y_2 + h^2 y_2(y_2^2 + r^2(x_2))^{-3/2} + y_3 = 0$$
$$f_3 = y_2 - 2y_3 + h^2 y_3(y_3^2 + r^2(x_3))^{-3/2} + y_4 = 0$$
$$\ldots$$
$$f_{39} = y_{38} - 2y_{39} + h^2 y_{39}(y_{39}^2 + r^2(x_{39}))^{-3/2} + y_{40} = 0$$

Here $y_0 = 0$ and $y_{40} = -0.0579$ are constants. As such, the corresponding system of linear equations following Newton's iterative method is as follows:

$$\begin{aligned}&-2u_1 + h^2(y_1^2 + r^2(x_1))^{-3/2}u_1\\ &+h^2 y_1(-3/2)(y_1^2 + r^2(x_1))^{-5/2}(2y_1u_1) + 1u_2\\ &= -(y_0 - 2y_1 + h^2 y_1(y_1^2 + r^2(x_1))^{-3/2} + y_2)\end{aligned}$$

$$\begin{aligned}&1u_1 - 2u_2 + h^2(y_2^2 + r^2(x_2))^{-3/2}u_2\\ &+h^2 y_2(-3/2)(y_2^2 + r^2(x_2))^{-5/2}(2y_2u_2) + 1u_3\\ &= -(y_1 - 2y_2 + h^2 y_2(y_2^2 + r^2(x_2))^{-3/2} + y_3)\end{aligned}$$

$$\begin{aligned}&1u_2 - 2u_3 + h^2(y_3^2 + r^2(x_3))^{-3/2}u_3\\ &+h^2 y_3(-3/2)(y_3^2 + r^2(x_3))^{-5/2}(2y_3u_3) + 1u_4\\ &= -(y_2 - 2y_3 + h^2 y_3(y_3^2 + r^2(x_3))^{-3/2} + y_4)\end{aligned}$$

$$\ldots$$

$$
\begin{aligned}
&1u_{38} - 2u_{39} + h^2(y_{39}^2 + r^2(x_{39}))^{-3/2}u_{39} \\
&+h^2 y_{39}(-3/2)(y_{39}^2 + r^2(x_{39}))^{-5/2}(2y_{39}u_{39}) + 1u_{40} \\
&= -(y_{38} - 2y_{39} + h^2 y_{39}(y_{39}^2 + r^2(x_{39}))^{-3/2} + y_{40})
\end{aligned}
$$

To solve these equations which are linear in u_i's, we have taken initial set of values of y[i]'s as

```
y[0]=0;
y[1]=0.1;y[2]=0.1;y[3]=0.1;y[4]=0.1;y[5]=0.1;y[6]
=0.1;y[7]=0.1;y[8]=0.1;y[9]=0.1;y[10]=0.1;y[11]=0.
1;y[12]=0.1;y[13]=0.1;y[14]=0.1;y[15]=0.1;y[16]=0.
1;y[17]=0.1;
y[18]=-0.1;y[19]=-0.1;y[20]=-0.1;y[21]=-0.1;y[22]=-
0.1;y[23]=-0.1;y[24]=-0.1;y[25]=-0.1;y[26]=-0.1;
y[27]=-0.1;y[28]=-0.1;
y[29]=0.1;y[30]=0.1;y[31]=0.1;y[32]=0.1;y[33]=0.1;y
[34]=0.1;y[35]=0.1;
y[36]=-0.1;y[37]=-0.1;y[38]=-0.1;y[39]=-0.1;
y[40]=-0.0579;
```

in program number 8.2. In using program number 8.2, the content of the program between the lines marked as

aaaaaaaaaaaaaaaaaaaaaaaaaaaaaaaaaaaaaa

and

bbbbbbbbbbbbbbbbbbbbbbbbbbbbbbbbbbbbb

should be pasted $p-1$ times, after the line

bbbbbbbbbbbbbbbbbbbbbbbbbbbbbbbbbbbbb

where p (= 5 in this case) is the number of iterations required to produce convergence of the values of y[i]'s. **The results of the program are shown in Table 8.2 and Figure 8.1.**

Program Number 8.2

```
c=0.2;
h=0.02;
y[0]=0;

y[40]=-0.0579;
y[1]=0.1;y[2]=0.1;y[3]=0.1;y[4]=0.1;y[5]=0.1;y[6]=
0.1;y[7]=0.1;y[8]=0.1;y[9]=0.1;y[10]=0.1;
```

Table 8.1

Euler Solution of Equation (8.1): $\frac{d^2y}{dx^2} = -\frac{y}{(y^2 + r^2(x))^{3/2}}$ with Initial Condition $(x, V) = (0, 1)$ and $(x, y) = (0, 0)$ where $\frac{dy}{dx} = V$

i	*X*	*V*	*y*
1	0.02	1.0000	0.0200
2	0.04	0.9506	0.0390
3	0.06	0.8577	0.0562
4	0.08	0.7308	0.0708
5	0.10	0.5798	0.0824
6	0.12	0.4127	0.0906
7	0.14	0.2355	0.0953
8	0.16	0.0523	0.0964
9	0.18	−0.1338	0.0937
10	0.20	−0.3202	0.0873
11	0.22	−0.5033	0.0772
12	0.24	−0.6776	0.0637
13	0.26	−0.8344	0.0470
14	0.28	−0.9614	0.0278
15	0.30	−1.0429	0.0069
16	0.32	−1.0644	−0.0144
17	0.34	−1.0187	−0.0347
18	0.36	−0.9098	−0.0529
19	0.38	−0.7501	−0.0679
20	0.40	−0.5540	−0.0790
21	0.42	−0.3338	−0.0857
22	0.44	−0.0979	−0.0877
23	0.46	0.1480	−0.0847
24	0.48	0.3993	−0.0767
25	0.50	0.6489	−0.0637
26	0.52	0.8841	−0.0461
27	0.54	1.0804	−0.0244
28	0.56	1.1996	−0.0005
29	0.58	1.2021	0.0236
30	0.60	1.0732	0.0451
31	0.62	0.8355	0.0618
32	0.64	0.5263	0.0723

(continued)

TABLE 8.1 (Continued)

i	*X*	*V*	*y*
33	0.66	0.1746	0.0758
34	0.68	−0.2029	0.0717
35	0.70	−0.5951	0.0598
36	0.72	−0.9811	0.0402
37	0.74	−1.3035	0.0141
38	0.76	−1.4422	−0.0147
39	0.78	−1.2851	−0.0404
40	0.80	−0.8757	−0.0579

```
y[11]=0.1;y[12]=0.1;y[13]=0.1;y[14]=0.1;y[15]=0.1;
y[16]=0.1;y[17]=0.1;
y[18]=-0.1;y[19]=-0.1;y[20]=-0.1;y[21]=-0.1;
y[22]=-0.1;y[23]=-0.1;y[24]=-0.1;y[25]=-0.1;
y[26]=-0.1;
y[27]=-0.1;y[28]=-0.1;
y[29]=0.1;y[30]=0.1;y[31]=0.1;y[32]=0.1;y[33]=0.1;
y[34]=0.1;y[35]=0.1;
y[36]=-0.1;y[37]=-0.1;y[38]=-0.1;y[39]=-0.1;
x=0;
i=0;
Table[{i=i+1,x=x+h,y[i]},{i,0,38,1}];
TableForm[%,TableSpacing->{3,3},TableHead-
ings->{None,{"i","x","y"}}]
x=0;
i=0;
p1=ListPlot[Table[{{i=i+1;x=x+h,
y[i]}},{i,0,38,1}],Frame->True,FrameLabel->
{"x","y"},
FrameTicks->All,PlotStyle->{Black},PlotRange->
Automatic]

aaaaaaaaaaaaaaaaaaaaaaaaaaaaa

G=NSolve[{
-2*u1+h^2*(y[1]^2+(-c*Cos[0.02]+0.5*c^2*(1-
1*Cos[2*0.02]))^2)^(-3/2)*u1+h^2*y[1]*(-
3/2)*(y[1]^2+
(-c*Cos[0.02]+0.5*c^2*(1-1*Cos[2*0.02]))^2)^(-
5/2)*(2*y[1]*u1)+1*u2==-(y[0]-
```

```
2*y[1]+h^2*y[1]*(y[1]^2+
(-c*Cos[0.02]+0.5*c^2*(1-1*Cos[2*0.02]))^2)^
(-3/2)+y[2]),
1*u1-2*u2+h^2*(y[2]^2+(-c*Cos[0.04]+0.5*c^2*
(1-1*Cos[2*0.04]))^2)^(-3/2)*u2+h^2*y[2]*
(-3/2)*(y[2]^2+(-c*Cos[0.04]+0.5*c^2*(1-
1*Cos[2*0.04]))^2)^(-5/2)*(2*y[2]*u2)+1*u3==
-(y[1]-2*y[2]+h^2*y[2]*(y[2]^2+(-c*
Cos[0.04]+0.5*c^2*(1-1*Cos[2*0.04]))^2)^
(-3/2)+y[3]),
1*u2-2*u3+h^2*(y[3]^2+(-c*Cos[0.06]+0.5*c^2*
(1-1*Cos[2*0.06]))^2)^(-3/2)*u3+h^2*y[3]*
(-3/2)*(y[3]^2+(-c*Cos[0.06]+0.5*c^2*(1-
1*Cos[2*0.06]))^2)^(-5/2)*(2*y[3]*u3)+1*u4==
-(y[2]-2*y[3]+h^2*y[3]*(y[3]^2+(-c*
Cos[0.06]+0.5*c^2*(1-1*Cos[2*0.06]))^2)^
(-3/2)+y[4]),
1*u3-2*u4+h^2*(y[4]^2+(-c*Cos[0.08]+0.5*c^2*
(1-1*Cos[2*0.08]))^2)^(-3/2)*u4+h^2*y[4]*
(-3/2)*(y[4]^2+(-c*Cos[0.08]+0.5*c^2*(1-
1*Cos[2*0.08]))^2)^(-5/2)*(2*y[4]*u4)+1*u5==
-(y[3]-2*y[4]+h^2*y[4]*(y[4]^2+(-c*
Cos[0.08]+0.5*c^2*(1-1*Cos[2*0.08]))^2)^
(-3/2)+y[5]),
1*u4-2*u5+h^2*(y[5]^2+(-c*Cos[0.10]+0.5*c^2*
(1-1*Cos[2*0.10]))^2)^(-3/2)*u5+h^2*y[5]*
(-3/2)*(y[5]^2+(-c*Cos[0.10]+0.5*c^2*(1-
1*Cos[2*0.10]))^2)^(-5/2)*(2*y[5]*u5)+1*u6==
-(y[4]-2*y[5]+h^2*y[5]*(y[5]^2+(-c*
Cos[0.10]+0.5*c^2*(1-1*Cos[2*0.10]))^2)^(-
3/2)+y[6]),
1*u5-2*u6+h^2*(y[6]^2+(-c*Cos[0.12]+0.5*c^2*
(1-1*Cos[2*0.12]))^2)^(-3/2)*u6+h^2*y[6]*
(-3/2)*(y[6]^2+(-c*Cos[0.12]+0.5*c^2*(1-
1*Cos[2*0.12]))^2)^(-5/2)*(2*y[6]*u6)+1*u7==
-(y[5]-2*y[6]+h^2*y[6]*(y[6]^2+(-c*
Cos[0.12]+0.5*c^2*(1-1*Cos[2*0.12]))^2)^
(-3/2)+y[7]),
1*u6-2*u7+h^2*(y[7]^2+(-c*Cos[0.14]+0.5*c^2*
(1-1*Cos[2*0.14]))^2)^(-3/2)*u7+h^2*y[7]*
(-3/2)*(y[7]^2+(-c*Cos[0.14]+0.5*c^2*
(1-1*Cos[2*0.14]))^2)^(-5/2)*(2*y[7]*u7)+1*u8==
-(y[6]-2*y[7]+h^2*y[7]*(y[7]^2+(-c*
Cos[0.14]+0.5*c^2*(1-1*Cos[2*0.14]))^2)^
(-3/2)+y[8]),
```

```
1*u7-2*u8+h^2*(y[8]^2+(-c*Cos[0.16]+0.5*c^2*
(1-1*Cos[2*0.16]))^2)^(-3/2)*u8+h^2*y[8]*
(-3/2)*(y[8]^2+(-c*Cos[0.16]+0.5*c^2*(1-
1*Cos[2*0.16]))^2)^(-5/2)*(2*y[8]*u8)+1*u9==
-(y[7]-2*y[8]+h^2*y[8]*(y[8]^2+(-c*
Cos[0.16]+0.5*c^2*(1-1*Cos[2*0.16]))^2)^
(-3/2)+y[9]),
1*u8-2*u9+h^2*(y[9]^2+(-c*Cos[0.18]+0.5*c^2*
(1-1*Cos[2*0.18]))^2)^(-3/2)*u9+h^2*y[9]*
(-3/2)*(y[9]^2+(-c*Cos[0.18]+0.5*c^2*(1-
1*Cos[2*0.18]))^2)^(-5/2)*(2*y[9]*u9)+1*u10==
-(y[8]-2*y[9]+h^2*y[9]*(y[9]^2+(-c*
Cos[0.18]+0.5*c^2*(1-1*Cos[2*0.18]))^2)^
(-3/2)+y[10]),
1*u9-2*u10+h^2*(y[10]^2+(-c*Cos[0.20]+0.5*c^2*
(1-1*Cos[2*0.20]))^2)^(-3/2)*u10+h^2*y[10]*
(-3/2)*(y[10]^2+(-c*Cos[0.20]+0.5*c^2*(1-
1*Cos[2*0.20]))^2)^(-5/2)*(2*y[10]*u10)+1*u11==
-(y[9]-2*y[10]+h^2*y[10]*(y[10]^2+(-c*
Cos[0.20]+0.5*c^2*(1-1*Cos[2*0.20]))^2)^
(-3/2)+y[11]),
1*u10-2*u11+h^2*(y[11]^2+(-c*Cos[0.22]+0.5*c^2*
(1-1*Cos[2*0.22]))^2)^(-3/2)*u11+h^2*y[11]*
(-3/2)*(y[11]^2+(-c*Cos[0.22]+0.5*c^2*(1-
1*Cos[2*0.22]))^2)^(-5/2)*(2*y[11]*u11)+1*u12==
-(y[10]-2*y[11]+h^2*y[11]*(y[11]^2+(-c*
Cos[0.22]+0.5*c^2*(1-1*Cos[2*0.22]))^2)^(-
3/2)+y[12]),
1*u11-2*u12+h^2*(y[12]^2+(-c*Cos[0.24]+0.5*c^2*
(1-1*Cos[2*0.24]))^2)^(-3/2)*u12+h^2*y[12]*
(-3/2)*(y[12]^2+(-c*Cos[0.24]+0.5*c^2*(1-
1*Cos[2*0.24]))^2)^(-5/2)*(2*y[12]*u12)+1*u13==
-(y[11]-2*y[12]+h^2*y[12]*(y[12]^2+(-c*
Cos[0.24]+0.5*c^2*(1-1*Cos[2*0.24]))^2)^(-
3/2)+y[13]),
1*u12-2*u13+h^2*(y[13]^2+(-c*Cos[0.26]+0.5*c^2*
(1-1*Cos[2*0.26]))^2)^(-3/2)*u13+h^2*y[13]*
(-3/2)*(y[13]^2+(-c*Cos[0.26]+0.5*c^2*(1-
1*Cos[2*0.26]))^2)^(-5/2)*(2*y[13]*u13)+1*u14==
-(y[12]-2*y[13]+h^2*y[13]*(y[13]^2+(-c*
Cos[0.26]+0.5*c^2*(1-1*Cos[2*0.26]))^2)^
(-3/2)+y[14]),
1*u13-2*u14+h^2*(y[14]^2+(-c*Cos[0.28]+0.5*c^2*
(1-1*Cos[2*0.28]))^2)^(-3/2)*u14+h^2*y[14]*
(-3/2)*(y[14]^2+(-c*Cos[0.28]+0.5*c^2*(1-
```

```
1*Cos[2*0.28]))^2)^(-5/2)*(2*y[14]*u14)+1*u15==
-(y[13]-2*y[14]+h^2*y[14]*(y[14]^2+(-c*
Cos[0.28]+0.5*c^2*(1-1*Cos[2*0.28]))^2)^(-
3/2)+y[15]),
1*u14-2*u15+h^2*(y[15]^2+(-c*Cos[0.30]+0.5*c^2*
(1-1*Cos[2*0.30]))^2)^(-3/2)*u15+h^2*y[15]*
(-3/2)*(y[15]^2+(-c*Cos[0.30]+0.5*c^2*(1-
1*Cos[2*0.30]))^2)^(-5/2)*(2*y[15]*u15)+1*u16==
-(y[14]-2*y[15]+h^2*y[15]*(y[15]^2+(-c*
Cos[0.30]+0.5*c^2*(1-1*Cos[2*0.30]))^2)^
(-3/2)+y[16]),
1*u15-2*u16+h^2*(y[16]^2+(-c*Cos[0.32]+0.5*c^2*
(1-1*Cos[2*0.32]))^2)^(-3/2)*u16+h^2*y[16]*
(-3/2)*(y[16]^2+(-c*Cos[0.32]+0.5*c^2*
(1-1*Cos[2*0.32]))^2)^(-5/2)*(2*y[16]*u16)+1*u17==
-(y[15]-2*y[16]+h^2*y[16]*(y[16]^2+(-c*
Cos[0.32]+0.5*c^2*(1-1*Cos[2*0.32]))^2)^
(-3/2)+y[17]),
1*u16-2*u17+h^2*(y[17]^2+(-c*Cos[0.34]+0.5*c^2*
(1-1*Cos[2*0.34]))^2)^(-3/2)*u17+h^2*y[17]*
(-3/2)*(y[17]^2+(-c*Cos[0.34]+0.5*c^2*(1-
1*Cos[2*0.34]))^2)^(-5/2)*(2*y[17]*u17)+1*u18==
-(y[16]-2*y[17]+h^2*y[17]*(y[17]^2+(-c*
Cos[0.34]+0.5*c^2*(1-1*Cos[2*0.34]))^2)^
(-3/2)+y[18]),
1*u17-2*u18+h^2*(y[18]^2+(-c*Cos[0.36]+0.5*c^2*
(1-1*Cos[2*0.36]))^2)^(-3/2)*u18+h^2*y[18]*
(-3/2)*(y[18]^2+(-c*Cos[0.36]+0.5*c^2*(1-
1*Cos[2*0.36]))^2)^(-5/2)*(2*y[18]*u18)+1*u19==
-(y[17]-2*y[18]+h^2*y[18]*(y[18]^2+(-c*
Cos[0.36]+0.5*c^2*(1-1*Cos[2*0.36]))^2)^
(-3/2)+y[19]),
1*u18-2*u19+h^2*(y[19]^2+(-c*Cos[0.38]+0.5*c^2*
(1-1*Cos[2*0.38]))^2)^(-3/2)*u19+h^2*y[19]*
(-3/2)*(y[19]^2+(-c*Cos[0.38]+0.5*c^2*(1-
1*Cos[2*0.38]))^2)^(-5/2)*(2*y[19]*u19)+1*u20==
-(y[18]-2*y[19]+h^2*y[19]*(y[19]^2+(-c*
Cos[0.38]+0.5*c^2*(1-1*Cos[2*0.38]))^2)^
(-3/2)+y[20]),
1*u19-2*u20+h^2*(y[20]^2+(-c*Cos[0.40]+0.5*c^2*
(1-1*Cos[2*0.40]))^2)^(-3/2)*u20+h^2*y[20]*
(-3/2)*(y[20]^2+(-c*Cos[0.40]+0.5*c^2*
(1-1*Cos[2*0.40]))^2)^(-5/2)*(2*y[20]*u20)+1*u21==
-(y[19]-2*y[20]+h^2*y[20]*(y[20]^2+(-c*
```

```
Cos[0.40]+0.5*c^2*(1-1*Cos[2*0.40]))^2)^
(-3/2)+y[21]),

1*u20-2*u21+h^2*(y[21]^2+(-c*Cos[0.42]+0.5*c^2*
(1-1*Cos[2*0.42]))^2)^(-3/2)*u21+h^2*y[21]*
(-3/2)*(y[21]^2+(-c*Cos[0.42]+0.5*c^2*(1-
1*Cos[2*0.42]))^2)^(-5/2)*(2*y[21]*u21)+1*u22==
-(y[20]-2*y[21]+h^2*y[21]*(y[21]^2+(-c*
Cos[0.42]+0.5*c^2*(1-1*Cos[2*0.42]))^2)^(-
3/2)+y[22]),
1*u21-2*u22+h^2*(y[22]^2+(-c*Cos[0.44]+0.5*c^2*
(1-1*Cos[2*0.44]))^2)^(-3/2)*u22+h^2*y[22]*
(-3/2)*(y[22]^2+(-c*Cos[0.44]+0.5*c^2*(1-
1*Cos[2*0.44]))^2)^(-5/2)*(2*y[22]*u22)+1*u23==
-(y[21]-2*y[22]+h^2*y[22]*(y[22]^2+(-c*
Cos[0.44]+0.5*c^2*(1-1*Cos[2*0.44]))^2)^
(-3/2)+y[23]),
1*u22-2*u23+h^2*(y[23]^2+(-c*Cos[0.46]+0.5*c^2*
(1-1*Cos[2*0.46]))^2)^(-3/2)*u23+h^2*y[23]*
(-3/2)*(y[23]^2+(-c*Cos[0.46]+0.5*c^2*(1-
1*Cos[2*0.46]))^2)^(-5/2)*(2*y[23]*u23)+1*u24==
-(y[22]-2*y[23]+h^2*y[23]*(y[23]^2+(-c*
Cos[0.46]+0.5*c^2*(1-1*Cos[2*0.46]))^2)^(-
3/2)+y[24]),
1*u23-2*u24+h^2*(y[24]^2+(-c*Cos[0.48]+0.5*c^2*
(1-1*Cos[2*0.48]))^2)^(-3/2)*u24+h^2*y[24]*
(-3/2)*(y[24]^2+(-c*Cos[0.48]+0.5*c^2*(1-
1*Cos[2*0.48]))^2)^(-5/2)*(2*y[24]*u24)+1*u25==
-(y[23]-2*y[24]+h^2*y[24]*(y[24]^2+(-c*
Cos[0.48]+0.5*c^2*(1-1*Cos[2*0.48]))^2)^(-
3/2)+y[25]),
1*u24-2*u25+h^2*(y[25]^2+(-c*Cos[0.50]+0.5*c^2*
(1-1*Cos[2*0.50]))^2)^(-3/2)*u25+h^2*y[25]*
(-3/2)*(y[25]^2+(-c*Cos[0.50]+0.5*c^2*
(1-1*Cos[2*0.50]))^2)^(-5/2)*(2*y[25]*u25)+1*u26==
-(y[24]-2*y[25]+h^2*y[25]*(y[25]^2+(-c*
Cos[0.50]+0.5*c^2*(1-1*Cos[2*0.50]))^2)^
(-3/2)+y[26]),
1*u25-2*u26+h^2*(y[26]^2+(-c*Cos[0.52]+0.5*c^2*
(1-2*Cos[1*0.52]))^2)^(-3/2)*u26+h^2*y[26]*
(-3/2)*(y[26]^2+(-c*Cos[0.52]+0.5*c^2*
(1-2*Cos[1*0.52]))^2)^(-5/2)*(2*y[26]*u26)+1*u27==
-(y[25]-2*y[26]+h^2*y[26]*(y[26]^2+(-c*
Cos[0.52]+0.5*c^2*(1-1*Cos[2*0.52]))^2)^
```

```
(-3/2)+y[27]),
1*u26-2*u27+h^2*(y[27]^2+(-c*Cos[0.54]+0.5*c^2*
(1-1*Cos[2*0.54]))^2)^(-3/2)*u27+h^2*y[27]*
(-3/2)*(y[27]^2+(-c*Cos[0.54]+0.5*c^2*
(1-1*Cos[2*0.54]))^2)^(-5/2)*(2*y[27]*u27)+1*u28==
-(y[26]-2*y[27]+h^2*y[27]*(y[27]^2+(-c*
Cos[0.54]+0.5*c^2*(1-1*Cos[2*0.54]))^2)^
(-3/2)+y[28]),
1*u27-2*u28+h^2*(y[28]^2+(-c*Cos[0.56]+0.5*c^2*
(1-1*Cos[2*0.56]))^2)^(-3/2)*u28+h^2*y[28]*
(-3/2)*(y[28]^2+(-c*Cos[0.560]+0.5*c^2*(1-
1*Cos[2*0.56]))^2)^(-5/2)*(2*y[28]*u28)+1*u29==
-(y[27]-2*y[28]+h^2*y[28]*(y[28]^2+(-c*
Cos[0.56]+0.5*c^2*(1-1*Cos[2*0.56]))^2)^
(-3/2)+y[29]),
1*u28-2*u29+h^2*(y[29]^2+(-c*Cos[0.58]+0.5*c^2*
(1-1*Cos[2*0.58]))^2)^(-3/2)*u29+h^2*y[29]*
(-3/2)*(y[29]^2+(-c*Cos[0.58]+0.5*c^2*
(1-1*Cos[2*0.58]))^2)^(-5/2)*(2*y[29]*u29)+1*u30==
-(y[28]-2*y[29]+h^2*y[29]*(y[29]^2+(-c*
Cos[0.58]+0.5*c^2*(1-1*Cos[2*0.58]))^2)^
(-3/2)+y[30]),
1*u29-2*u30+h^2*(y[30]^2+(-c*Cos[0.60]+0.5*c^2*
(1-1*Cos[2*0.60]))^2)^(-3/2)*u30+h^2*y[30]*
(-3/2)*(y[30]^2+(-c*Cos[0.60]+0.5*c^2*
(1-1*Cos[2*0.60]))^2)^(-5/2)*(2*y[30]*u30)+1*u31==
-(y[29]-2*y[30]+h^2*y[30]*(y[30]^2+(-c*
Cos[0.60]+0.5*c^2*(1-1*Cos[2*0.60]))^2)^
(-3/2)+y[31]),
1*u30-2*u31+h^2*(y[31]^2+(-c*Cos[0.62]+0.5*c^2*
(1-1*Cos[2*0.62]))^2)^(-3/2)*u31+h^2*y[31]*
(-3/2)*(y[31]^2+(-c*Cos[0.62]+0.5*c^2*
(1-1*Cos[2*0.62]))^2)^(-5/2)*(2*y[31]*u31)+1*u32==
-(y[30]-2*y[31]+h^2*y[31]*(y[31]^2+(-c*
Cos[0.62]+0.5*c^2*(1-1*Cos[2*0.62]))^2)^
(-3/2)+y[32]),
1*u31-2*u32+h^2*(y[32]^2+(-c*Cos[0.64]+0.5*c^2*(1-
1*Cos[2*0.64]))^2)^(-3/2)*u32+h^2*y[32]*
(-3/2)*(y[32]^2+(-c*Cos[0.64]+0.5*c^2*
(1-1*Cos[2*0.64]))^2)^(-5/2)*(2*y[32]*u32)+1*u33==
-(y[31]-2*y[32]+h^2*y[32]*(y[32]^2+(-c*
Cos[0.64]+0.5*c^2*(1-1*Cos[2*0.64]))^2)^
(-3/2)+y[33]),
1*u32-2*u33+h^2*(y[33]^2+(-c*Cos[0.66]+0.5*c^2*
```

```
(1-1*Cos[2*0.66]))^2)^(-3/2)*u33+h^2*y[33]*
(-3/2)*(y[33]^2+(-c*Cos[0.66]+0.5*c^2*
(1-1*Cos[2*0.66]))^2)^(-5/2)*(2*y[33]*u33)+1*u34==
-(y[32]-2*y[33]+h^2*y[33]*(y[33]^2+(-c*
Cos[0.66]+0.5*c^2*(1-1*Cos[2*0.66]))^2)^
(-3/2)+y[34]),
1*u33-2*u34+h^2*(y[34]^2+(-c*Cos[0.68]+0.5*c^2*
(1-1*Cos[2*0.68]))^2)^(-3/2)*u34+h^2*y[34]*
(-3/2)*(y[34]^2+(-c*Cos[0.68]+0.5*c^2*
(1-1*Cos[2*0.68]))^2)^(-5/2)*(2*y[34]*u34)+1*u35==
-(y[33]-2*y[34]+h^2*y[34]*(y[34]^2+(-c*
Cos[0.68]+0.5*c^2*(1-1*Cos[2*0.68]))^2)^
(-3/2)+y[35]),
1*u34-2*u35+h^2*(y[35]^2+(-c*Cos[0.70]+0.5*c^2*
(1-1*Cos[2*0.70]))^2)^(-3/2)*u35+h^2*y[35]*
(-3/2)*(y[35]^2+(-c*Cos[0.70]+0.5*c^2*
(1-1*Cos[2*0.70]))^2)^(-5/2)*(2*y[35]*u35)+1*u36==
-(y[34]-2*y[35]+h^2*y[35]*(y[35]^2+(-c*
Cos[0.70]+0.5*c^2*(1-1*Cos[2*0.70]))^2)^
(-3/2)+y[36]),
1*u35-2*u36+h^2*(y[36]^2+(-c*Cos[0.72]+0.5*c^2*
(1-1*Cos[2*0.72]))^2)^(-3/2)*u36+h^2*y[36]*
(-3/2)*(y[36]^2+(-c*Cos[0.72]+0.5*c^2*
(1-1*Cos[2*0.72]))^2)^(-5/2)*(2*y[36]*u36)+1*u37==
-(y[35]-2*y[36]+h^2*y[36]*(y[36]^2+(-c*
Cos[0.72]+0.5*c^2*(1-1*Cos[2*0.72]))^2)^
(-3/2)+y[37]),
1*u36-2*u37+h^2*(y[37]^2+(-c*Cos[0.74]+0.5*c^2*(1-
1*Cos[2*0.74]))^2)^(-3/2)*u37+h^2*y[37]*
(-3/2)*(y[37]^2+(-c*Cos[0.74]+0.5*c^2*
(1-1*Cos[2*0.74]))^2)^(-5/2)*(2*y[37]*u37)+1*u38==
-(y[36]-2*y[37]+h^2*y[37]*(y[37]^2+(-c*
Cos[0.74]+0.5*c^2*(1-1*Cos[2*0.74]))^2)^
(-3/2)+y[38]),
1*u37-2*u38+h^2*(y[38]^2+(-c*Cos[0.76]+0.5*c^2*
(1-1*Cos[2*0.76]))^2)^(-3/2)*u38+h^2*y[38]*
(-3/2)*(y[38]^2+(-c*Cos[0.76]+0.5*c^2*(1-
1*Cos[2*0.76]))^2)^(-5/2)*(2*y[38]*u38)+1*u39==
-(y[37]-2*y[38]+h^2*y[38]*(y[38]^2+(-c*
Cos[0.76]+0.5*c^2*(1-1*Cos[2*0.76]))^2)^
(-3/2)+y[39]),
1*u38-2*u39+h^2*(y[39]^2+(-c*Cos[0.78]+0.5*c^2*
(1-1*Cos[2*0.78]))^2)^(-3/2)*u39+h^2*y[39]*
(-3/2)*(y[39]^2+(-c*Cos[0.78]+0.5*c^2*
```

```
(1-1*Cos[2*0.78]))^2)^(-5/2)*(2*y[39]*u39)==
-(y[38]-2*y[39]+h^2*y[39]*(y[39]^2+(-c*
Cos[0.78]+0.5*c^2*(1-1*Cos[2*0.78]))^2)^
(-3/2)+y[40])},
{u1,u2,u3,u4,u5,u6,u7,u8,u9,u10,u11,u12,u13,u14,u15
,u16,u17,u18,u19,u20,
u21,u22,u23,u24,u25,u26,u27,u28,u29,u30,u31,u32,u33
,u34,u35,u36,u37,u38,u39}];

f[1]=N[EL1=Part[G,1];L1=u1/.EL1];
f[2]=N[EL2=Part[G,1];L2=u2/.EL2];
f[3]=N[EL3=Part[G,1];L3=u3/.EL3];
f[4]=N[EL4=Part[G,1];L4=u4/.EL4];
f[5]=N[EL5=Part[G,1];L5=u5/.EL5];
f[6]=N[EL6=Part[G,1];L6=u6/.EL6];
f[7]=N[EL7=Part[G,1];L7=u7/.EL7];
f[8]=N[EL8=Part[G,1];L8=u8/.EL8];
f[9]=N[EL9=Part[G,1];L9=u9/.EL9];
f[10]=N[EL10=Part[G,1];L10=u10/.EL10];
f[11]=N[EL11=Part[G,1];L11=u11/.EL11];
f[12]=N[EL12=Part[G,1];L12=u12/.EL12];
f[13]=N[EL13=Part[G,1];L13=u13/.EL13];
f[14]=N[EL14=Part[G,1];L14=u14/.EL14];
f[15]=N[EL15=Part[G,1];L15=u15/.EL15];
f[16]=N[EL16=Part[G,1];L16=u16/.EL16];
f[17]=N[EL17=Part[G,1];L17=u17/.EL17];
f[18]=N[EL18=Part[G,1];L18=u18/.EL18];
f[19]=N[EL19=Part[G,1];L19=u19/.EL19];
f[20]=N[EL20=Part[G,1];L20=u20/.EL20];
f[21]=N[EL21=Part[G,1];L21=u21/.EL21];
f[22]=N[EL22=Part[G,1];L22=u22/.EL22];
f[23]=N[EL23=Part[G,1];L23=u23/.EL23];
f[24]=N[EL24=Part[G,1];L24=u24/.EL24];
f[25]=N[EL25=Part[G,1];L25=u25/.EL25];
f[26]=N[EL26=Part[G,1];L26=u26/.EL26];
f[27]=N[EL27=Part[G,1];L27=u27/.EL27];
f[28]=N[EL28=Part[G,1];L28=u28/.EL28];
f[29]=N[EL29=Part[G,1];L29=u29/.EL29];
f[30]=N[EL30=Part[G,1];L30=u30/.EL30];
f[31]=N[EL31=Part[G,1];L31=u31/.EL31];
f[32]=N[EL32=Part[G,1];L32=u32/.EL32];
f[33]=N[EL33=Part[G,1];L33=u33/.EL33];
f[34]=N[EL34=Part[G,1];L34=u34/.EL34];
f[35]=N[EL35=Part[G,1];L35=u35/.EL35];
```

```
f[36]=N[EL36=Part[G,1];L36=u36/.EL36];
f[37]=N[EL37=Part[G,1];L37=u37/.EL37];
f[38]=N[EL38=Part[G,1];L38=u38/.EL38];
f[39]=N[EL39=Part[G,1];L39=u39/.EL39];
y[1]=f[1]+y[1];
y[2]=f[2]+y[2];
y[3]=f[3]+y[3];
y[4]=f[4]+y[4];y[5]=f[5]+y[5];
y[6]=f[6]+y[6];y[7]=f[7]+y[7];y[8]=f[8]+y[8];y[9]=
f[9]+y[9];y[10]=f[10]+y[10];
y[11]=f[11]+y[11];y[12]=f[12]+y[12];y[13]=f[13]+
y[13];y[14]=f[14]+y[14];y[15]=f[15]+y[15];
y[16]=f[16]+y[16];y[17]=f[17]+y[17];y[18]=f[18]+
y[18];y[19]=f[19]+y[19];y[20]=f[20]+y[20];
y[21]=f[21]+y[21];y[22]=f[22]+y[22];y[23]=f[23]+
y[23];y[24]=f[24]+y[24];y[25]=f[25]+y[25];
y[26]=f[26]+y[26];y[27]=f[27]+y[27];y[28]=f[28]+
y[28];y[29]=f[29]+y[29];y[30]=f[30]+y[30];
y[31]=f[31]+y[31];y[32]=f[32]+y[32];y[33]=f[33]+
y[33];y[34]=f[34]+y[34];y[35]=f[35]+y[35];
y[36]=f[36]+y[36];y[37]=f[37]+y[37];y[38]=f[38]+
y[38];y[39]=f[39]+y[39];
x=0;
i=0;
Table[{i=i+1,x=x+h,y[i]},{i,0,38,1}];
TableForm[%,TableSpacing->{3,3},TableHead-
ings->{None,{"i","x","y"}}]
x=0;
i=0;
p1=ListPlot[Table[{{i=i+1;x=x+h,
y[i]}},{i,0,38,1}],Frame->True,FrameLabel->
{"x","y"},
FrameTicks->All,PlotStyle->{Black},PlotRange->Auto-
matic]
```

bbbbbbbbbbbbbbbbbbbbbb

TABLE 8.2
The Results of the Program Number 8.2

i	x	Initial values of y_i's taken	y_i's after 5th iteration	Euler solution of Table 8.1
1	0.02	0.1	0.0220	0.0200
2	0.04	0.1	0.0429	0.0390
3	0.06	0.1	0.0618	0.0562
4	0.08	0.1	0.0780	0.0708
5	0.10	0.1	0.0909	0.0824
6	0.12	0.1	0.1003	0.0906
7	0.14	0.1	0.1060	0.0953
8	0.16	0.1	0.1078	0.0964
9	0.18	0.1	0.1058	0.0937
10	0.20	0.1	0.0999	0.0873
11	0.22	0.1	0.0901	0.0772
12	0.24	0.1	0.0766	0.0637
13	0.26	0.1	0.0596	0.0470
14	0.28	0.1	0.0396	0.0278
15	0.30	0.1	0.0174	0.0069
16	0.32	0.1	–0.0058	–0.0144
17	0.34	0.1	–0.0287	–0.0347
18	0.36	–0.1	–0.0496	–0.0529
19	0.38	–0.1	–0.0674	–0.0679
20	0.40	–0.1	–0.0811	–0.0790
21	0.42	–0.1	–0.0901	–0.0857
22	0.44	–0.1	–0.0942	–0.0877
23	0.46	–0.1	–0.0931	–0.0847
24	0.48	–0.1	–0.0867	–0.0767
25	0.50	–0.1	–0.0747	–0.0637
26	0.52	–0.1	–0.0572	–0.0461
27	0.54	–0.1	–0.0347	–0.0244
28	0.56	–0.1	–0.0082	–0.0005
29	0.58	0.1	0.0200	0.0236
30	0.60	0.1	0.0460	0.0451
31	0.62	0.1	0.0673	0.0618
32	0.64	0.1	0.0825	0.0723
33	0.66	0.1	0.0905	0.0758

(continued)

TABLE 8.2 (Continued)

i	x	Initial values of y_i's taken	y_i's after 5th iteration	Euler solution of Table 8.1
34	0.68	0.1	0.0909	0.0717
35	0.70	0.1	0.0831	0.0598
36	0.72	−0.1	0.0665	0.0402
37	0.74	−0.1	0.0409	0.0141
38	0.76	−0.1	0.0075	−0.0147
39	0.78	−0.1	−0.0282	−0.0404

Numerical solution to equation (8.1): $\frac{d^2y}{dx^2} = -\frac{y}{(y^2 + r^2(x))^{3/2}}$ with the boundary values $(x, y) = (0, 0)$ and $(0.8, -0.0579)$ using Newton's iterative method. Comparison with Euler solution with initial value $(x, y) = (0, 0)$ shows convergence to Euler solution.

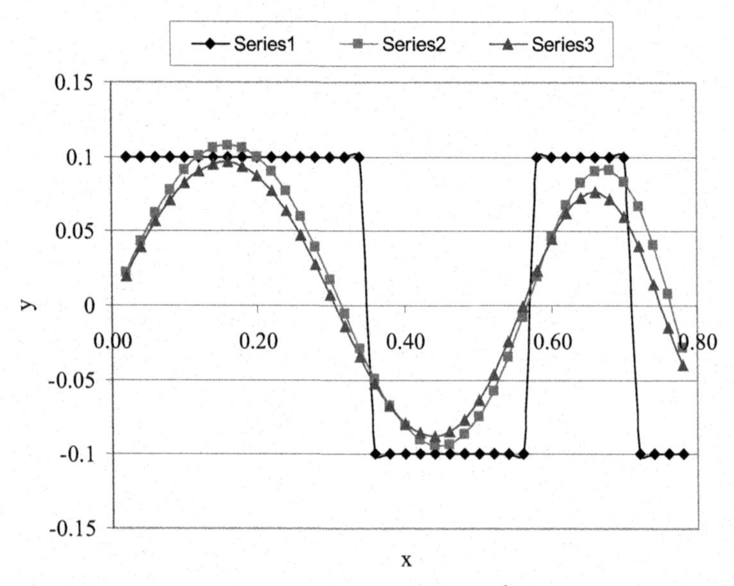

Figure 8.1 Numerical solution to equation (8.1): $\frac{d^2y}{dx^2} = -\frac{y}{(y^2 + r^2(x))^{3/2}}$ with the boundary values $(x, y) = (0, 0)$ and $(0.8, -0.0579)$ using Newton's iterative method. Showing the results of program number 8.2. Series 1 shows the initial values of y_i's taken, series 2 shows the results of Newton's method y_i's after 5th iteration, and series 3 shows the Euler solution of Table 8.1. Convergence to Euler solution is evident.

CONCLUDING REMARKS

1. The book presents the numerical solution to boundary value problems of a number of non-linear differential equations in comprehensive detail.
2. Replacing derivatives by finite difference approximations in these differential equations leads to a system of non-linear algebraic equations which we have solved using Newton's iterative method as applied to a system of non-linear algebraic equations.
3. In all cases, we have also obtained Euler solutions to the differential equations and ascertained that the iterations converge to Euler solutions.
4. Agreements between results of Newton's method and that of Euler's method served as a cross-check for any mistakes. No noteworthy differences were found.
5. We find that initial values of the 1st iteration need not be anything close to the final convergent values of the numerical solution.
6. The work explores the use of Newton's method in obtaining numerical solutions to boundary value problems of non-linear differential equations.
7. The book will contribute to applied mathematics and physics.

CONCLUDING REMARKS

REFERENCES

[1] Faruque, Syed Badiuzzaman, Solution of the Sitnikov Problem, *Celestial Mechanics and Dynamical Astronomy* **47** (2003): 353–369.

[2] Jacques, Ian, and Colin Judd, *Numerical Analysis*, Chapman and Hall, London, UK, 1987.

[3] Chowdhury, Sujaul, Ponkog Kumar Das, and Syed Badiuzzaman Faruque, *Numerical Solutions of Boundary Value Problems with Finite Difference Method*, IOP Concise Physics, Bristol, UK, 2018.

[4] Sastry, S. S., *Introductory Methods of Numerical Analysis* (5th Edition), PHI Learning, New Delhi, India, 2015.

[5] Sankara Rao, K., *Numerical Methods for Scientists and Engineers* (4th Edition), PHI Learning, New Delhi, India, 2017.

[6] Chowdhury, Sujaul, and Ponkog Kumar Das, *Numerical Solutions of Initial Value Problems Using Mathematica*, IOP Concise Physics, Bristol, UK, 2018.

INDEX

Note: Page numbers in *italics* indicate a figure and page numbers in **bold** indicate a table on the corresponding page.